Loïc Winterhalter

High throughput phenotyping of field-grown maize

Loïc Winterhalter

High throughput phenotyping of field-grown maize

Assessing important agronomical traits

Südwestdeutscher Verlag für Hochschulschriften

Impressum/Imprint (nur für Deutschland/only for Germany)
Bibliografische Information der Deutschen Nationalbibliothek: Die Deutsche Nationalbibliothek verzeichnet diese Publikation in der Deutschen Nationalbibliografie; detaillierte bibliografische Daten sind im Internet über http://dnb.d-nb.de abrufbar.
Alle in diesem Buch genannten Marken und Produktnamen unterliegen warenzeichen-, marken- oder patentrechtlichem Schutz bzw. sind Warenzeichen oder eingetragene Warenzeichen der jeweiligen Inhaber. Die Wiedergabe von Marken, Produktnamen, Gebrauchsnamen, Handelsnamen, Warenbezeichnungen u.s.w. in diesem Werk berechtigt auch ohne besondere Kennzeichnung nicht zu der Annahme, dass solche Namen im Sinne der Warenzeichen- und Markenschutzgesetzgebung als frei zu betrachten wären und daher von jedermann benutzt werden dürften.

Coverbild: www.ingimage.com

Verlag: Südwestdeutscher Verlag für Hochschulschriften GmbH & Co. KG
Heinrich-Böcking-Str. 6-8, 66121 Saarbrücken, Deutschland
Telefon +49 681 37 20 271-1, Telefax +49 681 37 20 271-0
Email: info@svh-verlag.de

Approved by: München, TU, Diss., 2011

Herstellung in Deutschland:
Schaltungsdienst Lange o.H.G., Berlin
Books on Demand GmbH, Norderstedt
Reha GmbH, Saarbrücken
Amazon Distribution GmbH, Leipzig
ISBN: 978-3-8381-2750-7

Imprint (only for USA, GB)
Bibliographic information published by the Deutsche Nationalbibliothek: The Deutsche Nationalbibliothek lists this publication in the Deutsche Nationalbibliografie; detailed bibliographic data are available in the Internet at http://dnb.d-nb.de.
Any brand names and product names mentioned in this book are subject to trademark, brand or patent protection and are trademarks or registered trademarks of their respective holders. The use of brand names, product names, common names, trade names, product descriptions etc. even without a particular marking in this works is in no way to be construed to mean that such names may be regarded as unrestricted in respect of trademark and brand protection legislation and could thus be used by anyone.

Cover image: www.ingimage.com

Publisher: Südwestdeutscher Verlag für Hochschulschriften GmbH & Co. KG
Heinrich-Böcking-Str. 6-8, 66121 Saarbrücken, Germany
Phone +49 681 37 20 271-1, Fax +49 681 37 20 271-0
Email: info@svh-verlag.de

Printed in the U.S.A.
Printed in the U.K. by (see last page)
ISBN: 978-3-8381-2750-7

Copyright © 2012 by the author and Südwestdeutscher Verlag für Hochschulschriften GmbH & Co. KG and licensors
All rights reserved. Saarbrücken 2012

Pluralitas non est ponenda sine neccesitate

(Plurality should not be posited without necessity)

William of Ockham

Or explaining Ockham's razor in other words:

The simplest explanation is most likely the correct one

Special thanks to ALL who made this possible!

Table of Contents

Zusammenfassung 7
Abstract 9
1. Outline of the Ph.D. thesis 11
   1.1 Publication I 11
   1.2 Publication II 12
   1.3 Publication III 13
2. Review of literature 15
   2.1 General introduction 15
   2.2 High throughput sensing 17
   2.3 Vertical profile of maize plants 19
   2.4 Spectral reflectance measurements 21
   2.5 Spectral indices 23
3. Objectives 25
4. Material and methods 27
5. Results and discussion 31
   5.1 Vertical footprint of maize canopies 31
   5.2 Aerial biomass and nitrogen uptake 32
   5.3 Canopy water mass and temperature 34
   5.4 Differentiation of temperate maize hybrids 36
   5.5 Difficulties and perspective 38
6. Conclusions 41
References 43
List of Publications 49
Appendix 50
Publication I
Publication II
Publication III

Zusammenfassung

Die zukünftige Klimaerwärmung und Wasserknappheit erfordern angepasste, trockenstressresistente Maispflanzen und den optimalen Einsatz knapper Wasserressourcen. Zielsetzung des Projekts war die Erfassung der Bestandeswassermenge und -temperatur, sowie der Biomasse und der Stickstoffaufnahme tropischer Maissorten (*Zea mays* L.) unter verschiedenen Trockenstressszenarien mittels nicht destruktiven Methoden. Bessere Kenntnisse auf diesem Gebiet können zur Optimierung von Managemententscheidungen beitragen, beispielweise zur standortangepassten Bewässerung und Düngung, aber auch in der Pflanzenzüchtung in der Selektion trockenstresstoleranter Sorten genutzt werden. Die Versuche wurden auf der Versuchsstation Dürnast der Technischen Universität München in den Jahren 2006-2007 und in Thailand auf der Versuchsstation des National Corn and Sorghum Research Center in den Jahren 2007-2009 durchgeführt. Regelmäßige Sensormessungen wurden während der Vegetationsperiode auf dem Feld durchgeführt, sowie begleitende Biomasseernten zur Erfassung der Bestandeswassermenge, der Biomasse und der Stickstoffaufnahme. Ausgewählte Spektralindizes sowie die IR-Temperatur wiesen eine hohe Korrelation mit der Bestandeswassermenge tropischer Maishybriden auf und es konnte gleichzeitig die Biomasse und die Stickstoffaufnahme durch Spektralindizes verlässlich erfasst werden, wobei die Kontroll- und Trockenstressvarianten differenziert werden konnten. Die optische Sensortechnologie, die auf einem Trägerfahrzeug angebracht war und mit GPS Daten kombiniert wurde, differenzierte deutlich verschiedene Stresslevels und ermöglichte eine Hochdurchsatz Präzisionserfassung der Bestandeswassermenge und -temperatur, sowie der Biomasse und der N-Aufnahme verschiedener Maishybriden. Die tropischen Maishybriden konnten sowohl mittels nicht destruktiven Sensormessungen, als auch basierend auf destruktiven Erhebungen konsistent in drei Gruppen unterteilt werden, unter-, durchschnittliche- und überdurchschnittliche Bestandeswassermenge, und besaßen ebenfalls ein konsistentes Ranking in der Biomasse und in der Stickstoffaufnahme bei den Kontroll- und Trockenstressvarianten. Der passive Reflexionssensor konnte die Biomasse und die Stickstoffaufnahme der Maisbestände bis zu den untersten Blattetagen erfassen und differenzierte dabei ebenfalls die unterschiedlichen Düngungsstufen. Die schräge und mehrfache Anordnung der Sensorgeometrie ermöglichte es einen großen „Fußabdruck" der einzelnen Parzellen zu erhalten und gleichzeitig den Bodeneinfluss auf die Sensormessungen zu reduzieren. Des Weiteren wurden die Blattflächenverteilung, die Biomasse und die Stickstoffaufnahme der einzelnen Blattetagen entlang der Maispflanzen untersucht, dabei wurde eine vertikale, glockenförmige Verteilung dargestellt. Die Ergebnisse zeigen, dass es möglich ist phänotypische und physiologische Bestandesmerkmale von Maissorten im Feldversuchswesen mit Hochdurchsatzsystemen zu erfassen, die sich als nützlich in der Optimierung von Managemententscheidungen wie auch in der Pflanzenzüchtung erweisen können.

Abstract

The future global warming and water shortage require adapted and drought tolerant maize plants, as well as an optimized input of scarce resources. Aim of the study was the detection of the canopy water mass and temperature, as well as the aerial biomass and nitrogen uptake of several maize hybrids (*Zea mays* L.) experiencing different drought stress treatments. A better knowledge in this area of expertise could improve screening for nitrogen uptake and drought tolerance in plant breeding programs and management decisions of farmers. The leaf levels of maize canopies have different contributions to spectral reflectance, detecting these would also improve the precision of canopy information. Experimental field trials were conducted at the research station Dürnast in Germany in the years 2006-2007 and at the National Corn and Sorghum Research Center in Thailand in the years 2007-2009. High throughput sensor measurements were performed regularly along with biomass samplings until flowering. Both spectral indices from literature and newly developed for this study were used for the evaluation. In our study we demonstrated that spectral indices as well as IR-temperature displayed a high correlation with canopy water mass and that spectral indices could also accurately estimate the aerial biomass and above ground nitrogen uptake of seven tropical maize hybrids grown under field conditions, while differentiating the drought stress levels. In addition, it was possible to classify consistently the hybrids examined in three groups (above, below or average performance) under control and stress environments with destructive and non-destructive measurements. The constant angle of view and an oblique and oligo viewing geometry allowed obtaining a large footprint of the maize plots while minimizing the soil influence in the field of view, showing that the passive reflectance sensor could differentiate the fertilization treatments and detect the aerial biomass and nitrogen uptake of the lowest leaf levels and furthermore demonstrating the vertical distribution of chlorophyll, aerial biomass, nitrogen content and uptake in maize canopies. Our results support the possibility of incorporating these methods in the development of non-destructive high throughput phenotyping techniques that could prove to be potentially useful for future plant breeding as well as improved nitrogen fertilization recommendations in crop management.

1. Outline of the Ph.D. thesis

The Ph.D. thesis consists of three parts, composed of **Publications I-III**, with the objectives of detecting agronomical traits of maize with high throughput precision phenotyping methods. The accurate positioning of the sensor and IR-thermometer with a constant angle of view and an oblique and oligo viewing geometry allowed obtaining a large footprint of the maize plots while minimizing the soil influence in the field of view. Reducing the time required for the measurements, being important for water status detection, further added to a new approach in estimating phenotypic and physiotypic traits of maize hybrids with carrier based, non-destructive, high throughput reflectance and thermal measurements co-recording GPS data. The experiments were conducted in Germany at the research station Dürnast for **Publication III** in the years 2006 – 2007 using maize at different fertilization rates and in Thailand for **Publication I** and **II** in the years 2007 – 2009, where seven hand sown tropical high yield hybrids were analyzed under four furrow irrigation treatments. Experimental field trials were carried out at the National Corn and Sorghum Research Center, where high throughput canopy reflectance and thermal radiance measurements using spectral indices from the literature and newly developed for this study, as well as biomass samplings were done regularly until flowering. The objective of **Publication III** was based on a fundamental research for this experiment, validating the efficacy of sensor measurements in determining the vertical leaf nitrogen uptake and aerial biomass of maize. The objectives of **Publication I** and **II** were to employ high throughput sensing measurements to detect the aerial biomass and nitrogen uptake, as well as to determine the canopy water mass of several tropical maize hybrids experiencing different drought stress treatments.

1.1 Publication I

The measurement of agronomical parameters of maize indicating its biomass and nutritional status provides important information to understand its responses to the environment. The detection of significant differences among maize hybrids would be very useful in plant breeding programs screening for nitrogen uptake and drought tolerance. The aim of **Publication I** was to assess the efficacy of high throughput sensing measurements in determining the aerial biomass and nitrogen uptake of tropical maize hybrids grown in well-watered and drought stress conditions. The relationship of the spectral indices with aerial biomass and nitrogen uptake had high coefficients of determination and also distinguished between drought stress levels. Through most sampling dates and irrigation levels, varieties were similarly classified in their amount of aerial biomass and nitrogen uptake by destructive and non-contacting measurements. Our results support the possibility of incorporating these methods in the development of high throughput phenotyping techniques potentially useful for future plant breeding.

1.2 Publication II

The high throughput determination of the water status of maize in precision agriculture presents numerous benefits, but also shows the potential for improvement. On the one hand, the differentiation of maize hybrids could be used in screening drought tolerance in plant breeding, while on the other hand, monitoring the plant status with carrier based sensors could enable a fast evaluation of various traits over a large area, therefore improving management decisions of farmers. The aim of **Publication II** was to assess the ability to measure the canopy water mass (CWM; amount of water in kg m^{-2}) of several tropical maize hybrids using high throughput sensing. Spectral indices from literature and newly developed were validated. Selected spectral indices and IR-temperature were highly correlated with CWM and differentiated the drought stress levels. Several indices had in average high coefficients of determination, while being able to differentiate and classify the hybrids into three consistent groups (above, below, or average performance) under control and stress environments. The results of this study show that it is indeed possible to both detect CWM and discriminate between groups of hybrids using non-destructive high throughput phenotyping, demonstrating that this technology presents a potentially useful application for breeding in the future.

1.3 Publication III

The different leaf levels of plant canopies have different contributions to spectral reflectance, in detecting these, this would improve the precision of canopy information. The aim of **Publication III** was to validate the efficacy of sensor measurements to determine the vertical leaf nitrogen uptake and aerial biomass of maize. The questions were how nitrogen is distributed in the plants and do differences exist among fertilization levels, as well as how deep into the canopy a passive reflectance based sensor reaches: does it detect the nitrogen uptake of the whole canopy or only the upper leaves?. A concave distribution of the relative chlorophyll content in maize plants was shown, while the vertical nitrogen content had steeper gradients from top to bottom with lower fertilization rates. Leaf nitrogen uptake distribution in maize resembled the bell shape distribution of aerial biomass, except for showing a clear differentiation of the fertilization rates in the middle leaf levels. The sensor detected the nitrogen uptake of each leaf level, even the lowest levels, whereas the spectral index R_{780}/R_{740} was strongly and curvilinearly related ($R^2=1.00$) to the nitrogen uptake of maize resulting from the alternating removal of leaf levels starting from the bottom leaves followed by reflectance measurements. Although more than half of the total nitrogen in maize was stored in the stem, the index values were mainly influenced by the foliage. Our results support the possibility to incorporate this information in improved nitrogen fertilization recommendations in crop management as well as in precision phenotyping, also demonstrating that the information of the vertical nitrogen profile could be utilized in crop growth simulation models used for research.

2. Review of literature

2.1 General introduction

The race between increasing world population and food production has become a fierce competition, since drought as one of the major limitations to food production worldwide is likely to worsen in the future with the predicted global climate change, therefore maintaining high yields under drought conditions has become a priority (Berger et al., 2010; Serraj et al., 2011). Drought stress limits plant productivity as well as crop yields by reducing photosynthesis and leaf growth (Hunt et al., 1987). Periods of soil water deficit can arise at any time during the growing season, causing in the tropics average annual yield losses of about 17% for maize (*Zea mays* L.), but also reached 60% in individual seasons in regions such as southern Africa (Edmeades et al., 1999). The worldwide annual harvest yield of maize was 784 million tons in 2007, obviously higher than the harvest quantities of rice (*Oryza sativa* L.) with 651 million tons and wheat (*Triticum aestivum* L.) with 607 million tons (FAO, 2010). As a consequence of this, the cultivation of maize is important for global food security and significant yield losses from drought in key traditional production areas are expected to increase (Campos et al., 2004).

However, the physiological basis of reaching high yields under drought conditions remains barely understood due to the various mechanisms that plants can use to maintain growth in conditions of low water supply and the complexity of the stress itself (Berger et al., 2010). Furthermore, the interaction of water and nitrogen stress has a wide influence on the yield of crops, since nitrogen uptake is dependent on soil and plant water status (Hu and Schmidhalter, 2005). As a consequence, nitrogen should be applied to areas with sufficient plant available water, to avoid costly and suboptimal nitrogen applications on water-limited crops, optimizing nitrogen management, also improving remote sensing based nitrogen fertilization recommendations (Clay et al., 2006; Tilling et al., 2007), so that farmers could profit from remote sensing techniques to assist in management decisions (Osborne et al., 2002). Within a given field, spatial differences in soil characteristics will cause plant development and nitrogen demand to vary (Diker and Bausch, 2003). Assessing these variations in plant development is important, particularly when they are caused by stress conditions that can be easily corrected through management practices that employ rapid and precise measurements of factors that hinder crop growth and limit biomass production (Poss et al., 2006). Heterogeneous plant canopies need locally adapted nitrogen fertilizer applications based on fast and accurate measurements of the local N status (Mistele and Schmidhalter, 2008). Using remote sensing data to help in such agronomic decisions provides a method to improve field-scale management (Samborski et al., 2009) by varying nitrogen application rates according to spatially variable crop nitrogen requirements to avoid over- or under-fertilization (Thoren and Schmidhalter, 2009), resulting in an optimized fertilizer management with economical and ecological advantages. Besides, the detection of water stress with sensors in real-time, can be applied in precision irrigation systems, leading to a more efficient water use and optimizing the photosynthesis as well as plant primary productivity (Claudio et al., 2006). Sensor systems based on measurements of leaf reflectance, or canopy temperature, could automatically trigger the irrigation improving the irrigation scheduling through a more accurate timing and before a visual drought stress is noticeable, negatively affecting the yield (Seelig et al., 2008a).

Present methods to determine the plant water status are time-consuming, e.g. measuring water potential by pressure chamber, soil water content or leaf water content measurements, and require various observations to characterize a field. Managers could profit from remote sensing techniques to assist in irrigation management decisions (Graeff et al., 2007). Fertilization management is commonly based on soil analysis, but the shortcoming of this time-consuming method is the lack of taking into account any local variation in nitrogen demand or mineralization (Mistele and Schmidhalter, 2008). Additionally, the direct analysis of plant material can be employed as indicator for the nitrogen supply from the soil within the growing period, or a non-destructive handheld device for chlorophyll detection can been used. The chlorophyll meter SPAD 502 (Minolta Corp., Japan) is a small device measuring the light transmittance at the red wavelength (650 nm) for chlorophyll absorption and near-infrared (960 nm) to correct the chlorophyll measurement for leaf thickness (Olfs et al., 2005). Linear relationships between the SPAD values and chlorophyll content of leaves were achieved, while the results also indicated that the SPAD could be successfully applied in crop nitrogen management. However, the limitations lie in the difficulty to efficiently gather chlorophyll information over a large area quickly from SPAD values (Huang et al., 2010).

2.2 High throughput sensing

Precision farming needs carrier based sensors for collecting crop information from agricultural fields that are quick, precise, efficient and non-destructive, as well as simple-to-use and cost-effective. Implementing such methods allows a fast detection of the spatial and temporal variability of the soil water and nitrogen supply influencing the biomass production and nitrogen uptake of crops, leading to possible applications in the site-specific crop investigation and nutrient management (Reusch et al., 2002; Schmidhalter et al., 2006). Adapting the nitrogen supply to the water availability both spatially and temporally is crucial to reach the requirements for an optimal crop growth, especially early in the season when management decisions can affect the yield, offering opportunities for precision crop nitrogen management (Tilling et al., 2007). The use of remote sensing data to support agronomic decisions improving field-scale management might have a significant impact on yield and profitability, since a moderate temporary stress or deficiency of plant growth factors already limit the biomass of crops (Hatfield et al., 2008). Detecting differences in the nitrogen status of crops within a field with a carrier based sensor, while applying the right amount of nitrogen fertilizer according to the spatial variable and actual crop requirements, has positive effects through avoiding over- or under-fertilization (Lammel et al., 2001). Therefore spatially targeted and temporally optimized nitrogen fertilizer applications are very welcome for economical as well as ecological reasons (Schmidhalter et al., 2006), since nitrogen as an important nutrient for modern crop production is often over-applied without considering the crop requirements or potential environmental risk to ensure high crop yields (Hatfield et al., 2008).

Also in the domain of precision and site-specific irrigation the insufficiency of real-time soil or plant status feedback and decision support systems is a major limitation, while precision agriculture has the potential to become the next key improvement to the use-efficiency of crop-production inputs (Peters and Evett, 2007). Given, that progress in the real-time detection of water stress with sensors has been made, this could be applied in precision irrigation systems leading to a more efficient water use, optimized photosynthesis and plant primary productivity (Claudio et al., 2006). Sensor systems based on measurements of the plant water status through the canopy temperature by using infrared thermometry (Hunt and Rock, 1989) or leaf reflectance, could automatically trigger the irrigation leading to an improved irrigation scheduling due to a more accurate timing, before even a visual drought stress is ascertainable and consequently the yield would be negatively affected (Osborne et al., 2002; Seelig et al., 2008a). Higher canopy temperatures for drought stressed plants are the result of a decreased stomatal conductance leading to a less latent heat loss from transpiration and an increase in leaf temperature, due to this fact an array of temperature sensors mounted on irrigation platforms could be regularly used to assess drought stress on fields creating canopy temperature maps to provide real-time information on the crop water status (Hunt and Rock, 1989; Peters and Evett, 2007). In summary, the detection of the nitrogen, biomass and water status of crops by contact less sensor measurements is seen to be a promising technique not only for management decisions of farmers but also for breeding purposes (Schmidhalter, 2005).

The common practice in plant breeding is to visually score plant traits of genotypes in field trials as an indirect measure of tolerance to stresses (cold, heat, drought, etc.), while this method has its advantages, however, this is a subjective and not a quantitative measure exposed to human error. As a result of the insufficiency of adequate techniques for non-destructive phenotyping of the above-ground biomass of many genotypes in field trials, breeding, genomic, and physiological research on early growth in plants is restricted (Montes et al., 2011). A major breakthrough in the development of high throughput, high-precision phenotyping systems under drought is based on the development of repeatable and predictive field screening methods suitable for use in breeding programs, allowing genes for yield components under stress to be efficiently mapped and their effects assessed on a range of drought-related traits (Serraj et al., 2011). In order to include these techniques into molecular breeding programs and identifying the underlying genes, reliable phenotyping protocols are necessary. While high throughput methods have made gene and marker identification very fast, the phenotyping process still limits the progress in genetics, demonstrating that the development of high throughput phenotyping technologies are essential to close the gap between plant physiology and genetics, this combination being particularly important for studies of drought tolerance, since the complex reactions of plants to drought require a dissection of the responses into a series of component traits (Berger et al., 2010). Since plant traits that improve water uptake, water use efficiency and partitioning to yield work synergistically to maximize productivity under water-limited environments, precision phenotyping could improve the discovery of these genes and understanding the complex interactions among genes, genetic background and the environment (Reynolds et al., 2009). The identification of genes in breeding programs that affect performance under drought stress leading to improved varieties depend on the development of simple, repeatable, low-cost, high throughput phenotyping screening methods that trustworthy characterize genetic variation for drought resistance and its component traits (Serraj et al., 2011). Canopy temperature is measured using infrared thermometry and is ideal for high throughput screening because it is quick and easy to measure, while the technology is inexpensive and shows good relationships with performance. Integrating these measurements into the breeding process showed that selecting for cooler plots additionally to visual selection for plant type, improved the ability to identify high yielding lines (Reynolds et al., 2009). Also, the high throughput non-destructive biomass determination possesses a great potential to advance research towards to a better understanding of the genetical, physiological and biochemical basis of plant growth and enable in plant breeding a large scale evaluation of germplasm for selection of desirable genotypes under field conditions. While improvements in chip technology enabled cheap high throughput genotyping, phenotyping of these traits in field trials is still a major bottleneck, therefore high throughput phenotyping platforms are urgently needed to fully utilize the potential of genomic tools (Montes et al., 2011). High throughput approaches in precision phenotyping such as infrared thermometry or spectral reflectance have the potential to be a cost effective way of increasing genetic gains in breeding populations (Reynolds et al., 2009).

2.3 Vertical profile of maize plants

The vertical insight or footprint of sensors is largely unknown. The vertical leaf layers of plant canopies have different contributions to spectral reflectance. Identifying the contributions of different layers on the reflectance could improve the precision of canopy information gained through remote sensing data (Wang et al., 2005), leading to a better understanding of sensor measurements for high throughput phenotyping techniques. In literature little information is available that characterizes the distribution of biomass, chlorophyll and nitrogen in maize, describing the relationship between spectral reflectance information and the distribution of these parameters. The knowledge of such relationships could improve the assessment of biomass as well as optimize management decisions and also enhance the understanding of plant phenotypes and could serve as a basis for architectural plant modeling. It is essential to understand the nitrogen uptake and assimilation to improve the nitrogen use efficiency of crops by adjusting nitrogen fertilizer applications (Gastal and Lemaire, 2002) and therefore improving nitrogen management (Huang et al., 2010). There are strong correlations between the photosynthetic activity and nitrogen content of leaves, since 75% of the leaf nitrogen is integrated in the photosynthetic process (Drouet and Bonhomme, 1999). The leaf chlorophyll concentration is mainly determined by the availability of nitrogen, therefore the precise assessment of the chlorophyll status of plants is essential to provide nitrogen fertilization recommendations (Filella et al., 1995). Relationships between chlorophyll content of leaves and actual photosynthetic canopy area were reported in literature, but almost no information is available about the vertical distribution of chlorophyll as important crop biophysical characteristic (Ciganda et al., 2008). A non-uniform nitrogen distribution is based on the fact that leaves of a plant canopy are exposed to different light environments have a difference in age and may also develop under varying nitrogen supply situations during growth, while leaf appearance remains constant (Gastal and Lemaire, 2002). During plant development the local light climate plays a crucial role and has a significant influence on the leaf nitrogen distribution and remobilization, since the vertical gradient of leaf nitrogen content per unit area is dependent on the vertical gradient in leaf irradiance during the vegetative phase of maize (Drouet and Bonhomme, 1999). The vertical chlorophyll content distribution, as well as the area-per-leaf profile of maize canopies can be characterized in every growth stage with a slightly skewed bell-shaped curve, with the possibility of being biologically interpreted and consequently analyzing changes in the profile inherent to growing seasons and agronomic practices (Ciganda et al., 2008; Keating and Wafula, 1992; Valentinuz and Tollenaar, 2006).

The analysis of the vertical leaf area profile of maize canopies could improve the accurate estimation of the radiation interception and canopy photosynthesis for crop growth simulation models computing dry matter accumulation from temporal integration of canopy photosynthesis (Valentinuz and Tollenaar, 2006). Furthermore, the quantification of the chlorophyll content in plant canopies could complement the information of the leaf area index, improving the comprehension of the crop ecophysiology, the interplant competition, the radiation use efficiency and its productivity (Ciganda et al., 2008). Multiple studies predicting canopy photosynthesis based the model calculations on measurements of the nitrogen content of selected leaves (Wang et al., 2005), rarely considering the nitrogen partitioning during maize development in crop models. Therefore to model precisely dry matter production and grain yield, the analysis of changes in nitrogen gradients in relation to changes in the local light climate within heterogeneous row crops (e.g. maize) should be incorporated (Drouet and Bonhomme, 1999). The accuracy of leaf area predictions affects the performance of crop growth models used for research and management, with the bell-shaped function being a source for improvement, since leaf area is an important determining factor for crop growth (Keating and Wafula, 1992). The accuracy of models depends on the realism with which the plant canopies are represented and the possibility of employing a priori knowledge on canopy properties to restrain the inversion procedure and although models simulate canopy reflectance mostly well, their application is limited to the simulation of homogeneous canopies and could lead to significant errors and falsification when applied to row crops and additionally it would be useful to investigate the sensitivity in the inversion of different models to the well known saturation effect that occurs with dense canopy covers where reflectance becomes insensitive to changes in the leaf area index since the lower layers of foliage are not visible, all things considered leading to improvements in the realism of models used for the inversion with remotely sensed data and to more accurate estimations of crop biophysical characteristics with the use of realistic and accurate input variables (Casa et al., 2010).

2.4 Spectral reflectance measurements

Factors indicating the water and nutritional status of leaves are commonly used to explain different physiological and ecological phenomena or problems, giving essential information about the plants, but obtaining the dynamics of an identical leaf is difficult, since measurements are commonly done through destructive methods (Yu et al., 2000). A tractor-based oblique quadrilateral-view optic reflectance sensor offers a fast and non-destructive determination of the crop nitrogen status (Mistele and Schmidhalter, 2008). Reflectance is mostly characterized by differences in the optical densities between the saturated cell walls and the intercellular region (Gates et al., 1965). Variations in leaf cell structure and composition induced through drought stress influence the spectral reflectance values, because of changes in the properties of connections between cell walls and air spaces, cell sizes and shapes, and cell wall composition and structure (Grant, 1987; Liu et al., 2003; Penuelas et al., 1994). Reflection occurs mainly in the layer of spongy mesophyll cells as well as in the layer of palisade cells due to differences between cell walls and intercellular air spaces (Seelig et al., 2008b). The plant water status directly influences the intercellular air spaces and the cell turgidity, consequently changing the leaf cell structure that affects how light will be absorbed, transmitted, or reflected by leaves (Schlemmer et al., 2005; Seelig et. al, 2008b), influencing leaf spectral reflectance. The quantity of light absorbed by leaves is a function of its photosynthetic pigment content, thus the chlorophyll content directly determines the photosynthetic potential and primary production, while providing an indirect estimation of the nutrient status, since the leaf enzyme Rubisco as main sink for nitrogen is connected to the chlorophyll (Chapelle et al., 1984). Changes of canopy reflectance are the largest in the near-infrared wavelengths during the growing season due to the increase of biomass, while the visible wavelength range also shows significant seasonal variations relating to the absorption of light by photosynthetic pigments (Hatfield et al., 2008). Reflectance in the red edge wavelength region contains information about chlorophyll absorption, the cell wall reflection and additionally the alteration between these main effects. The absorbance is still high although the reflectance at 700 nm is beyond the maximum absorbance of chlorophyll and the beginning of the red edge, showing that an increased nitrogen content influences both the intensity of the reflection and the inflection point in the red edge (Mistele and Schmidhalter, 2008). Furthermore, using the reflectance in the near infrared gives an effective estimation of the plant water status (Yu et al., 2000). The 740 – 800 nm wavelength spectrum is strongly influenced by the leaf cell structure (Schlemmer et al., 2005) and the strongest relationships between leaf reflectance and chlorophyll content are observed in the green spectrum near 550 mm and the far red spectrum near 700 nm (Gitelson et al. 2003; Mistele and Schmidhalter, 2010), while the spectral reflectance in the red edge (680 – 740 nm) and near infrared (740 – 940 nm) (Liu et al., 2003; Tilling et al., 2007) is also influenced by changes in the leaf internal structure caused by changes in the water status of leaves. Drought stressed leaves tend to have an increased reflectance in the 400 - 1300 nm region (Graeff and Claupein, 2007; Inoue et al., 1993).

The reflectance signals from plant canopies are a function of the leaf spectral and morphological properties, the variation of leaf cover over the soil, the architectural arrangement of the leaves, branches and stems, percentage vegetative cover, and the atmospheric signals between the surface and the sensor (Hatfield et al., 2008). Although basic correlations with parameters could be established, the practical applicability is still difficult, since the results may be influenced by extraneous variables, such as sun and sensor viewing angle, illumination intensity, scene surface heterogeneity, background properties, and atmospheric optical variability (Seelig et al., 2008b). In the atmosphere water vapor is absorbed, meaning that the energy reaching the ground surface is greatly reduced enhancing the background influence, leading to problems in field trials when trying to estimate the water content of whole plant canopies (Rollin and Milton, 1989). Using a multiple direction viewing geometry reduces the azimuth effects, but the reflectance spectra is still affected by the solar zenith angle due to the non-Lambertian reflectance of the canopy (Reusch, 2003). A number of extraneous variables influence reflectance-based indices as well and should therefore been considered (Seelig et al., 2008a).

2.5 Spectral indices

The development of sensor systems enable a high throughput screening approach for comparing spectral reflectance indices of genotypes, since the composition of light reflected by plant canopies is a function of several physiological factors including light interception, hydration status of leaf tissues, pigment content and composition of photosynthetic tissue (Reynolds et al., 2009). A major challenge for researchers is to fully understand the potential of remote sensing as a source of information used for agronomic management decisions, which requires an expansion of the basic knowledge of the agronomic information content of remote sensing data, and algorithmic innovations for analysis (Hatfield et al., 2008). Vegetation indices have been developed to relate reflectance of plant canopies with canopy characteristics to estimate agriculturally important plant parameters. Among the most commonly used spectral indices are two-wavelength ratio vegetation indices (Diker and Bausch, 2003; Gutierrez et al., 2010; Reusch, 2003). These indices can give good predictions of total aerial dry matter and aboveground N-uptake (Mistele and Schmidhalter, 2010), as well as plant water content (Claudio et al., 2006; Penuelas et al., 1997) and relative water content (Inoue et al., 1993; Yu et al., 2000). Spectral indices derived from real-time multispectral reflectance information also have the potential to be used as indicators of plant water status and deficit stress (Claudio et al., 2006; Seelig et al., 2009). The near-infrared spectrum has several water absorption bands that offer in combination the possibility to measure the leaf water content. Leaf water indices can be calculated by dividing reflectance at a spectral region that is only weakly absorbed by water by that at a region that is strongly absorbed by water, thereby measuring the absolute amount of absorbing water that is present within the path of light reflected from leaves directly, showing that the index R1300/R1450 could detect the onset of leaf dehydration non-destructively and in real-time (Seelig et al., 2009). A good choice of wavelengths for plant canopies are placed in the local green maximum (550 nm), in the red edge (670 - 750 nm) and NIR (> 750 nm) region, which provide for good predictions of aboveground N-uptake of crops (Reusch et al., 2002; Reusch, 2003). The correlation of reflectance indices to crop parameters demonstrates a strong relationship between indices such as REIP or NIR/NIR and total aerial N (Mistele and Schmidhalter, 2008).

3. Objectives

The main objective of this Ph.D. thesis was partitioned in three segments, represented by **Publications I-III**.
The objective of **Publication III** was based on a fundamental research for this experiment. In this work a method was used that differentiates and integrates the spectral information received vertically along whole maize canopies and the relationship to biomass and nitrogen contained in the individual leaves from top to bottom of the canopy. The aim was to validate sensor measurements to determine the vertical leaf nitrogen uptake of maize. The questions were how nitrogen is distributed in maize plants and do differences exist between different fertilization rates. Also, how does a passive reflectance sensor see maize canopies, does it detect the nitrogen uptake of the whole plant or only the upper leaves, meaning how deep does the sensor reach into the canopy. This would lead to a better understanding of the information gained from spectral measurements of plant canopies for precision farming and phenotyping. Fertilizer decisions are frequently based on reference values obtained only from selected point measurements and therefore may not fully reflect the nitrogen distribution, in contrast to spectral reflectance measurements of plant canopies giving more accurate information about the nitrogen uptake of maize plants in a specific area. A more holistic approach would also be to consider the nitrogen of stem.
The objectives of the other two publications were to employ high throughput sensing measurements to detect agronomical traits of maize, meaning that in **Publication I** the aim was to detect the aerial biomass and nitrogen uptake, while in **Publication II** the aim was to determine the canopy water mass of several tropical maize hybrids experiencing different drought stress treatments. Drought stress is a difficult trait to evaluate because of the unpredictable occurrence of natural droughts and lack of information on effective screening techniques (Serraj et al., 2011), therefore the experiments were conducted during the dry season in Thailand.
The combination of high throughput phenotyping joined with high throughput genotyping would certainly lead to a major breakthrough in understanding the fundamental genetic basis of complex plant traits like drought tolerance (Edmeades et al., 2004). Unfortunately, our ability to characterize gene-to-phenotype relationships for drought tolerance in maize is hampered by the delay in developing

methods to efficiently measure plant phenotypes for important traits, since fast advances have been made in plant genotyping. Data obtained from high throughput precision phenotyping, would give the researcher detailed information of plant phenotypes determined by various genotypes from a plant breeding population with the possibility of understanding this relationship (Campos et al., 2004; Thoren and Schmidhalter, 2009). However, it is important to realize that the relationship of spectral reflectance and nitrogen status is different between crop varieties. To compensate for this effect, reference values are necessary to determine the fertilizer application rate (Lammel et al., 2001), which in turn could be used for differentiating maize varieties in plant breeding. The most important restriction in improving selection methods for drought stress environments is the deficit in fast and precise measurements of the plant phenotype, while enhancing the speed and precision of phenotyping plant traits would be more effective in the progress of plant breeding than additional advances in molecular technologies. The evaluation of large populations in plant breeding requires that early selection and effective reduction be conducted as quickly and as cheaply as possible (Richards et al., 2010).

4. Material and methods

The experiments were carried out in Germany and Thailand. The trials in the years 2006 to 2007 were conducted on the research station Dürnast (11.70 E long; 48.40 N lat; 450 m alt) located near Freising in Germany in the tertiary hills of the Bavarian Alps. The research station Dürnast belongs to the Chair of Plant Nutrition at the Technical University of Munich. This location has an average annual precipitation of 800 mm and temperature of 7.5 °C, with an average temperature from April to September of 13.3 °C and a sunshine duration of 118 3 h during this period. Three maize field experiments were carried out and cultivation methods were done following local technical recommendations. The trials in Thailand were conducted at the National Corn and Sorghum Research Center (101.3 E long; 14.6 N lat; 380 m alt), located 155 km northeast of Bangkok. The climate in this tropical region is influenced by a rainy and dry season, the latter occurring from November to February and being characterized by no rainfall. Trials were carried out in the dry seasons in the years 2007/08 and 2008/09. The average temperature (November 1 to January 31) in the first year was 24 °C and 22.5 °C in the second year. The soil of the research station is classified as reddish brown lateritic soil with moderately high water permeability (Neidhart, 1994). Seven maize hybrids, four irrigation treatments and four replications were arranged as a randomized block design. Cultivation methods (e.g. thinning and hand weeding) were done following local technical recommendations and there was no need to do any pest management.

The sensor system with modified electronics (tec5, Oberursel, Germany) contained two units of a Zeiss MMS1 silicon diode array spectrometer that measure reflectance and incident radiation at the same time. The optics were mounted with an angle of 50° on the corners of the frame, leading to an oblique and oligo view optic. One component was connected to a four-in-one light fiber to create an optical mixed signal of the canopy reflection of one area from four different directions, with the spectrometer analyzing the reflected radiation in 256 spectral channels with a detection range from 300 to 1000 nm and a bandwidth of 3.3 nm. The second component was connected to a diffuser to measure the global radiation to compensate for different light conditions, with the associated spectrometer having a detection range from 1000 to 1700 nm and a bandwidth of 6 nm (Mistele and Schmidhalter, 2008).

For the experiment in Germany the passive reflectance sensor was mounted on a high clearance tractor (BRAUD 2714) used as platform with a measuring height of 3 m (above ground) that allowed to measure in fully developed maize canopies. Canopy reflectance measurements as well as biomass samplings were done alternating, while succeedingly removing hierarchical higher leaf levels from bottom to top in an area of 28 m^2 representing the sensor's field of view. At first, the original entire maize canopy was measured, then 1-2 leaf levels and in the end the stalks were removed while doing sensor reflectance measurements. Canopy reflectance measurements in Thailand were done until flowering, as simultaneously to the biomass samplings as possible and always at the same time (between 11:00 and 13:00). The sensor system was mounted on a forklift, with the frame being placed on the left side of the carrier, measuring exactly over the maize canopy and with the possibility to vary the height of the sensor enabling measurements of fully developed maize canopies with plant heights of 2.5 m. Driving lanes next to the plots allowed a fast and non-destructive possibility to measure the maize canopies through the entire growing period. Biomass harvests in Thailand were performed regularly until flowering to determine aerial biomass and nitrogen uptake, as well as the amount of water in kg ha^{-1} contained in the above ground biomass (= CWM) of the maize plants. At each harvest, 30 plants were cut from 6 m in the inner rows (not border plants) and their fresh weight was determined. After determining the fresh weight of the biomass samplings (Germany and Thailand), the plant material was chopped, with a representative subsample being collected and weighed before being oven dried at 100 °C for 3 days and reweighed. The dried samples were ball milled (100 µm) and analyzed for total nitrogen content with an Isotope Radio Mass Spectrometer (IRMS) combined with a preparation unit (ANCA SL 20-20, Europe Scientific, Crewe, UK).

The canopy temperature (Thailand) was measured with an infrared (IR) thermometer (KT15D, Heitronics Infrarot Messtechnik GmbH, Wiesbaden, Germany) that was also mounted on the sensor system so that the measurements were done simultaneously with those of canopy reflectance. All spectral and thermal radiance measurements were co-registered with corresponding GPS data. Additionally, tensiometers were installed at depths of 20, 40, 60, 80 and 100 cm to monitor the soil matric potential and to control the drying out of the soil during the drought stress periods. All measurements were done regularly until flowering. Additionally, the chlorophyll content of maize leaf levels from bottom to top was measured with the chlorophyll meter SPAD 502 for the experiment in Germany.

Statistical analysis was conducted to determine the effect of the maize hybrid and different irrigation treatments, both individually and in combination, on the relationship of aerial biomass, nitrogen uptake and CWM with the different spectral indices. The spectral index R_{780}/R_{740}, developed as optimized index (Mistele and Schmidhalter, 2008; Winterhalter et al., 2011), was used for the experiment in Germany to establish and allow comparisons to agronomic traits of maize canopies. Curvilinear models (quadratic) in Microsoft Excel 2003 (Microsoft Inc., Seattle, WA, USA), as well as analysis of variance (ANOVA) and general linear model (GLM) analysis in SPSS 16 (SPSS Inc., Chicago, USA) were used to establish relationships. Finally, hybrids were classified as having above, below or average performance using the method of Worku et al. (2007).

5. Results and discussion

5.1 Vertical footprint of maize canopies

The SPAD measurements showed a concave distribution of the relative chlorophyll content in maize plants, with the highest SPAD values at or above the middle leaf levels and differentiating the nitrogen fertilizer applications. This is similar to the results of Ciganda et al. (2008), who reported that the vertical chlorophyll content distribution of maize was best described through a bell-shaped curve, while in contrast to winter wheat a vertical chlorophyll distribution with a decreasing trend from the top to the bottom was found (Huang et al., 2010). The nitrogen content was similarly distributed along the different leaf levels with slightly lower values at the bottom leaf levels. The literature reported that a vertical gradient in the leaf nitrogen concentration is common in crop canopies (Wang et al., 2005). Also in the vegetative phase of maize a vertical gradient in leaf nitrogen content per unit area was shown (Drouet and Bonhomme, 1999). The aerial biomass in maize plants displayed a vertical bell shaped function with the highest values towards the middle leaf levels, while the leaf nitrogen uptake distribution closely resembled the bell shape distribution of the aerial biomass, except for a clearer differentiation of the fertilization rates in the middle leaf levels, leading to higher values with increased fertilization, as presented in **Publication III**. There is no information available about this matter in the literature, apart from the area-per-leaf profile of maize canopies, which were described with a slightly skewed bell-shaped function (Keating and Wafula, 1992; Valentinuz and Tollenaar, 2006). This information could improve nitrogen management and nitrogen use efficiency through an advanced understanding of the nitrogen uptake and assimilation (Gastal and Lemaire, 2002), while also improving the performance of crop growth models used in research and management (Keating and Wafula, 1992; Valentinuz and Tollenaar, 2006), since the nitrogen partitioning during maize development is rarely considered in crop growth models (Drouet and Bonhomme, 1999). The spectral index R_{780}/R_{740} used in previous studies to relate agronomical traits of maize canopies with spectral reflectance (Mistele and Schmidhalter, 2008; Winterhalter et al., 2011) was closely and curvilinearly related ($R^2=1.00$) to the aerial biomass and the nitrogen of maize resulting from the alternating removal of leaf levels starting from the bottom leaves followed by reflectance measurements. With each removal of a leaf level the index values, the aerial biomass and the nitrogen uptake values decreased, even at the lowest leaf levels, although a small saturation effect was observed indicating that the sensor underestimated the aerial biomass and nitrogen uptake of the bottom leaf levels. Casa et al. (2010) reported that reflectance is insensitive to changes in the leaf area index, because lower layers of maize canopies are not at all visible. While small saturation effects were detected especially with high fertilization rates during the late vegetation period, however, the information is useful for improvements in high-throughput phenotyping in plant breeding (Winterhalter et al., 2011). The sensor system could certainly detect accurately the nitrogen uptake in earlier growing stages, supporting on-the go fertilizer applications (Mistele and Schmidhalter, 2010; Schmidhalter et al., 2003), since an increased fertilization led to higher index values, differentiating the fertilization levels. Even though more than half of the total nitrogen in maize was stored in the stem, the index values were mainly influenced by the foliage. Reflectance values from the remaining stems or the bare soil were very similar. This is the first report on the impact of maize stem on the proportion of the reflectance of whole plant canopies and also no differentiation of foliage and stem reflection for maize was discovered in the literature. A more accurate estimation of crop biophysical characteristics could be gained through improvements in the realism of models used for the inversion with remotely sensed data (Casa et al., 2010).

5.2 Aerial biomass and nitrogen uptake

Irrigation treatments as well as tropical maize hybrids had a significant influence on the aerial biomass and nitrogen uptake. The longer maize plants experienced drought stress leading to less available water, the lower the aerial biomass and nitrogen uptake was, showing that the availability of water obviously is essential for plant growth and development. Maize plants subjected to drought stress have a lower leaf-area index, plant height, and biomass accumulation than under irrigated conditions (Edmeades et al., 1999; Soler et al., 2007). The spectral reflectance of maize canopies is influenced by the chlorophyll content and N status of plants (Hatfield et al., 2008; Mistele und Schmidhalter, 2008; Schlemmer et al., 2005), as well as drought stress at different growth stages (Clay et al., 2006). The results demonstrated that several spectral indices showed strong relationships during the entire vegetative stage and could successfully detect the significant influence of the different irrigation treatments on the aerial biomass and nitrogen uptake of the tropical maize hybrids. Similar results were reported by Mistele and Schmidhalter (2008) for temperate maize, showing strong relationships of the index NIR/NIR (R_{780}/R_{740}) with total aerial N, whereas Diker and Bausch (2003) also showed that vegetation indices correlate well with dry matter at different growth stages. For example the relationships of aerial biomass with the spectral index ($R_{790}-R_{720}$)/($R_{790}+R_{720}$), as well as nitrogen uptake with the spectral index R_{760}/R_{730} had high coefficients of determination, while showing clear differences between the treatments and accurately assessing the aerial biomass and nitrogen uptake of maize canopies under different drought stress levels. An overview of the quality for the relationship of aerial biomass and nitrogen uptake with spectral indices from literature as well as the newly developed for this study is presented in **Publication I**. The tropical maize hybrids showed differences and were similarly classified in their amount of aerial biomass and nitrogen uptake. A statistically based grouping of the hybrids was possible, while significant differences in aerial biomass and nitrogen uptake were found only for hybrids with the maximum and minimum values, which agrees with the results of Edmeades et al. (1999), who found that differences in the biomass production of tropical maize experiencing drought and in well-watered environments were significant only for cultivars with the greatest and the lowest total aboveground biomass. All varieties are commercial high yield hybrids planted in this region of Thailand and have passed a wide area testing for several years before being released to commercial uses, consequently leading to small performance differences among the hybrids. The aim was to detect even little differences between the genotypes and if the differences were consistent, a ranking of the hybrids would be possible and could potentially improve the breeding process. Soler et al. (2007) reported that different maize hybrids respond differently to soil water limitations, showing that genotypic differences in the response to drought stress, whereas Inoue et al. (1993) stated that the differentiation of maize cultivars is important in plant breeding for screening tests of drought tolerant plants. The advantage of our method is to deliver high throughput measurements of plant parameters on a large area, given that the sensor was mounted on a forklift, demonstrating that the spectral information obtained by a tractor-mounted crop scanning device allows for a fast detection of the biomass and N status of plants on-the-go, important for improving precision farming techniques (Schmidhalter et al., 2001) and likely for precision phenotyping.

5.3 Canopy water mass and temperature

Tensiometers were used to control the drying out of the soil with increasing drought stress, also indicating the major rooting depth and water uptake of the maize plants, given that a moderate decrease in soil matric potential led to a severe drought stress because of the low available soil water capacity being typical for this experimental site (Camp, 1996; Neidhart, 1994). Measurements of the soil water status were important to back up sensor measurements of the drought stressed hybrids. Irrigation treatments and tropical maize hybrids had individually a significant influence on the CWM for almost all measurements in the GLM analysis for the experimental years 2007/08 and 2008/09. Hybrids were similarly classified in their amount of CWM for all sampling dates and irrigation treatments and a statistically based grouping of the hybrids was possible, comparable to the results of the aerial biomass and nitrogen uptake, also showing that significant differences in CWM were found only for hybrids with the maximum and minimum values. The different drought stress treatments could be distinguished through the height of the index values, showing that higher index values were achieved with increased irrigation. Therefore, the index values decreased with increasing drought stress created through withholding irrigation during a certain period, which agrees with the results of Yu et al. (2000), who reported that drought stress impacts on the physiological properties of maize, leading to a decrease in the values of physiological indices. The spectral indices could indeed detect the significant influence of different irrigation treatments on the CWM of maize. The best indices (R_{850}/R_{725}; R_{780}/R_{740}; R_{760}/R_{730}; R_{890}/R_{715}; $(R_{790}-R_{720})/(R_{790}+R_{720})$; R_{980}/R_{715}) had strong relationships with CWM during the vegetative growing period. Hunt and Rock (1989) reported that vegetation indices utilizing red and NIR wavelengths could be employed to detect plant water stress, especially the total leaf water mass per unit ground area, or the water absorption peak near 970 nm like the WBI (Penuelas et al., 1994). The excellent ability of the newly developed spectral index R_{850}/R_{725} to assess the water status of maize canopies was confirmed through the good relationship with CWM ($R^2 = 0.83$). An overview of the relationship between the CWM and all spectral indices tested is presented in **Publication II**, with the six best indices having coefficients of determination (R^2) exceeding 0.70 over all experiments from 2007 to 2009. It is also possible to determine drought stress through infrared thermometry (Tilling et al., 2007). IR-temperature had strong relationships with CWM for all measurements and could distinguish between the different stress treatments of the canopy, showing that an increased water availability lead to a lower canopy temperature, which agrees with the results of Hunt and Rock (1989) and Peters and Evett (2007), who reported higher canopy temperatures for drought stressed plants. Measurements of canopy temperature detecting spatial variations in the water status could improve management decisions by adjusting the inputs to the heterogeneous soil and growing conditions within a field, leading to a decrease in input costs while reducing the pollution of the environment (Osborne et al., 2002; Tilling et al., 2007). To detect the water status of maize plants through non-destructive, high-throughput sensing measurements, IR-temperature was related e.g. to the spectral index R_{850}/R_{725}, showing a strong relationship and a clear differentiation of the irrigation treatments. The tropical maize hybrids were classified into three consistent groups (above, below or average performance) under control and stress environments for the investigated vegetative growing period. The ability to discriminate between maize cultivars will improve screening methods of drought tolerance in plant breeding (Inoue et al., 1993). Genotypic differences exist in the vegetative responses to drought stress, since maize plants respond differently to a restricted water supply, therefore management strategies should be improved like using cultivars best adapted to local environmental conditions and optimal irrigation strategies (Soler et al., 2007). Although stronger relationships between spectral indices and CWM were present compared to aerial biomass and nitrogen uptake, it shows a strong relationship between CWM and biomass demonstrating the connection of these important agronomical traits. Our method allows for a new approach in plant breeding assessing phenotypic and physiotypic traits of maize hybrids with carrier based, non-destructive, high throughput reflectance and thermal measurements with corresponding GPS data.

5.4 Differentiation of temperate maize hybrids

The leaf nitrogen uptake distribution of twelve maize hybrids formed a vertical bell shaped function with the highest values being achieved in the middle leaf levels, also showing differences between the cultivars (Figure 1). Similar results were reported from Keating and Wafula (1992), as well as Valentinuz and Tollenaar (2006) for the area-per-leaf profile of maize canopies, which could be described through a slightly skewed bell-shaped function. The spectral index R_{780}/R_{740} was highly and curvilinearly related (R^2=1.00) to the nitrogen uptake of the twelve maize hybrids resulting from the concomitant removal of leaf levels starting from the bottom leaves, and followed by sensor measurements (Figure 2). With each removal of leaf levels the index values and the nitrogen uptake values decreased, even at the lowest leaf levels, similarly to the results of **Publication III** for the different fertilization treatments. The results do not indicate a clear differentiation of the hybrids, which could possibly be explained through the influence of the time of day and zenith angle on the sensor measurements. The removal of leaf levels was labor intensive causing the experiment to last from 11:00 to 16:00 o'clock on five days, showing differences between sensor measurements at midday, afternoon and morning. The maize hybrids stored approximately half of the nitrogen uptake in the foliage and half in the stem, leading to a leaf/stem ratio of about 1 for the twelve hybrids (Table 3).

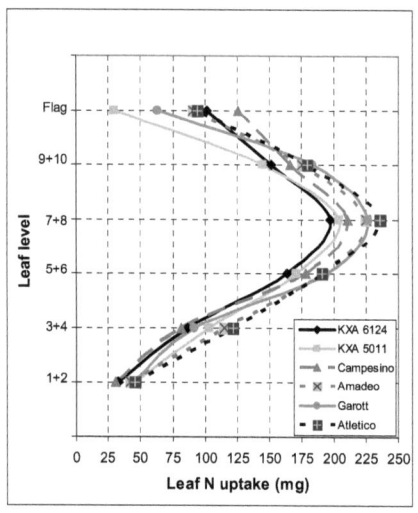
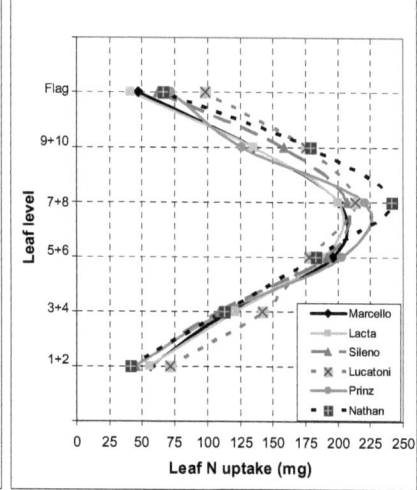

Figure 1. Vertical nitrogen uptake profiles of twelve maize hybrids in 2007.

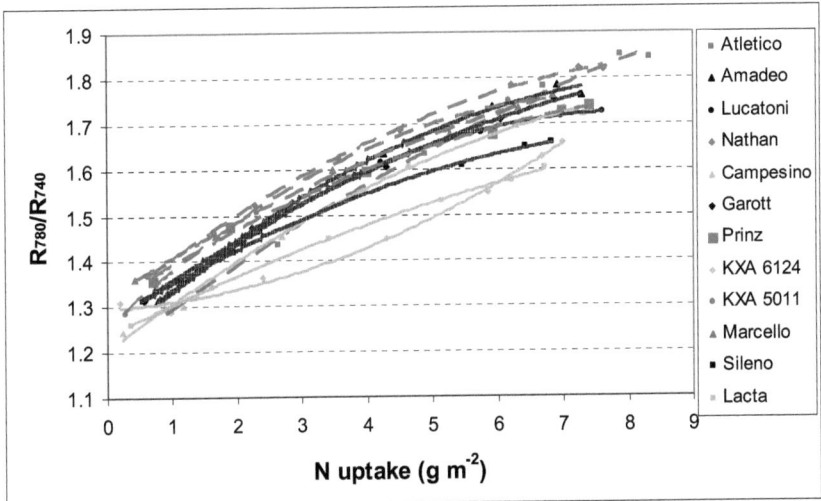

Figure 2. Relationship of the spectral index R_{780}/R_{740} with nitrogen uptake of maize canopies (twelve hybrids, $R^2 = 1.00$) resulting from the concomitant removal of succeedingly higher leaf levels and immediately following sensor measurements in 2007. Segmented lines represent sensor measurements at midday, dark lines in the afternoon and continuous lines in the morning.

Table 1. Nitrogen uptake of the foliage and the stem, as well as the leaf/stem ratio of twelve maize hybrids in 2007.

Hybrid	N uptake of foliage (mg/plant)	N uptake of stem (mg/plant)	Leaf/stem ratio of N uptake
KXA 6124	732	784	0.9
KXA 5011	697	891	0.8
Garott	803	703	1.1
Campesino	793	767	1.0
Atletico	868	865	1.0
Amadeo	840	875	1.0
Marcello	752	793	0.9
Lucatoni	880	729	1.2
Lacta	743	591	1.3
Nathan	824	857	1.0
Prinz	779	634	1.2
Sileno	774	747	1.0

5.5 Difficulties and perspective

For the experiments in Germany the differentiation of maize hybrids through the vertical footprint was hampered by the influence of the time of day and zenith angle. Even though strong relationships of index values with aerial biomass or nitrogen uptake were present in Thailand, the hybrids could not directly been differentiated, probably caused by the high variability between replications, inhibiting the detection of differences among the hybrids. This variation might be due in part to the local agricultural standard (e.g., hand sowing and fertilizing) and/or the variability of the soil. The high average coefficients of variation of 18 % for aerial biomass and 20 % for nitrogen uptake underline this suggestion. Under best possible cultivation conditions coefficient of variations may be close to 10% (Schmidt, 1995). To counteract these problems, further experiments on the research station Dürnast are carried out under better conditions and higher agricultural standards (technological improvements, e.g. mechanization of the cultivation methods) in 2011 with 17 maize hybrids, also conducting sensor measurements of the vertical footprint at the same time of day, enabled through an improved experimental design, leading to an even more precise differentiation of the hybrids under controlled conditions. Also the lack of correlation between CWM and grain yield could be caused by the high coefficient of variation, covering the small differences between the hybrids. Furthermore, the long recovery period between the last drought stressed biomass sampling and the final harvest could explain the lack of correlation. Maize suffering from drought stress in the vegetative stage has a high recovery ability, also shown at the same site by Camp (1996) and Neidhart (1994). The results demonstrated that a lower CWM at the vegetative stage is not necessarily related to final yield, but plants with a higher CWM could probably have a deeper rooting system and therefore might be able to survive a longer drought stress period. The results are still highly interesting because they show phenotype differences among the hybrids, this being potentially useful in further screening processes. Whether the individual recovery ability of hybrids was higher or other agronomical traits may have contributed to this observation must be investigated in follow-up studies at later growth stages detailing further vegetative and reproductive growth parameters. In addition, inducing drought stress for a longer phase or rather closer to the harvest, or even several drought stress periods until the harvest may be a way of solving this problem. Differentiating vegetative growth parameters at early growth stages are the foundation to analyze also later growth stages, with the potential to find close links that may exist between spectrally determined biomass parameters in vegetative growth stages and the biomass yield of silage and energy maize. If the already demonstrated potential to discriminate hybrids can be extended to later growth stages as well, then a deeper understanding of traits contributing to the overall performance and to the final yield of drought stressed maize may become available.

6. Conclusions

The vertical distribution of chlorophyll, aerial biomass, nitrogen content and uptake in maize canopies was described, while the passive reflectance sensor could differentiate the fertilization treatments as well as the hybrids and detect the aerial biomass and nitrogen uptake of the lowest leaf levels with a slight underestimation, demonstrating the vertical footprint of maize canopies. The spectral indices could accurately determine the aerial biomass and N uptake of several tropical maize hybrids at different drought stress levels, with the hybrids being mostly similarly classified in their amount of aerial biomass, N uptake and index values under well watered and drought stressed environments during the vegetative growing period. Spectral indices as well as IR-temperature displayed a high correlation with CWM and could also differentiate between plants experiencing the drought stress levels. Furthermore, the maize hybrids could be classified consistently in three groups (above, below or average performance) under control and stress environments. Overall, our results show that it is possible to detect the aerial biomass and N uptake, as well as the CWM of maize hybrids at different irrigation treatments and to discriminate between groups of hybrids using high throughput phenotyping. Although there was no consistent relationship with grain yield, this method still has potential to be used in the selection process of silage or energy maize cultivars. A main improvement in sensing methods represented by our technique is mounting the sensor on a carrier and not using handheld devices, enabling high-throughput measurements of plant parameters on a large area. The exact positioning of the reflectance sensor and IR-thermometer with a constant angle of view and an oblique and oligo viewing geometry allowed obtaining a large footprint of the maize plots while minimizing the soil influence in the field of view. Researchers in the field of precision farming that employ tractor based sensors commonly utilize a nadir view geometry, bringing forth sensor heights of 9 m for maize, due to the need of at least 2.5 rows in the field of view (Major et al., 2003). The advantage of our oblique and oligo viewing geometry is the significant reduction of the sensor height above the maize canopy, while detecting 5 rows. Furthermore, reducing the time required for the measurements, which is very important for the water status detection, adds to a new approach in plant breeding estimating phenotypic and physiotypic traits of maize hybrids with carrier based, non-destructive, high throughput precision reflectance and thermal measurements with combined GPS data. Our results reveal the strength and weakness of passive reflectance sensor measurements, while supporting the possibility to incorporate this information in improved nitrogen fertilization recommendations in crop management. In addition, the information gained of the vertical nitrogen profile could be utilized in crop growth simulation models used for research. However, although high clearance tractors have been employed in precision farming, it is new to analyze and differentiate maize hybrids in plant breeding through the combination of high throughput precision phenotyping with sensors mounted on a carrier, while mapping the information with GPS data. High throughput precision phenotyping coupled with high throughput genotyping has the potential to give researchers the ability to obtain detailed information of plant genotype characteristics influencing the phenotype, as well as the possibility of directly selecting the phenotype. A combined approach could lead to a major breakthrough in the attempts to explain the fundamental genetic basis of drought tolerance in maize (Campos et al., 2004; Richards et al., 2010).

References

Berger, B., Parent, B., Tester, M., 2010. High-throughput shoot imaging to study drought responses. J. Exp. Bot. doi:10.1093/jxb/erq201.

Camp, K.H., 1996. Transpiration efficiency of tropical maize (Zea Mays L.). Ph.D. diss. Swiss Federal Institute of Technology Zürich, Switzerland.

Campos, H., Cooper, M., Habben, J.E., Edmeades, G.O.., Schussler, J.R., 2004. Improving drought tolerance in maize: A view from industry. Field Crop Res. 90:19–34.

Casa, R., Baret, F., Buis, S., Lopez-Lozano, R., Pascucci, S., Palombo, A., Jones, H.G., 2010. Estimation of maize canopy properties from remote sensing by inversion of 1-D and 4-D models. Precision Agric DOI 10.1007/s11119-010-9162-9.

Chappelle, E., McMurtrey III, J., Wood, F., Newcomb. W.,1984. Laser-induced fluorescence of green plants. 2. LIF caused by nutrient deficiencies in corn. Appl. Optics. 23: 139–142.

Hunt, E. R., Jr., Rock, B. N., Nobel, P. S., 1987. Measurement of leaf relative water content by infrared reflectance, Remote Sens. Environ. 22:429–435.

Ciganda, V., Gitelson, A., Schepers, J., 2008. Vertical profile and temporal variation of chlorophyll in maize canopy: Quantitative "Crop Vigor" indicator by means of reflectance-based techniques. Agron. J. 100:1409–1417.

Claudio, H.C., Cheng, Y., Fuentes, D.A., Gamon, J.A., Luo, H., Oechel, W., Qiu, H.L., Rahman, A.F., Sims, D.A., 2006. Monitoring drought effects on vegetation water content and fluxes in chaparral with the 970 nm water band index. Remote Sens. Environ. 103:304–311.

Clay, D., Kim, K., Chang, J., Clay, S., Dalsted K., 2006. Characterizing water and nitrogen stress in corn using remote sensing. Agron. J. 98:579–587.

Edmeades, G.O., Bolanos, J., Chapman, S.C., Lafitte, H.R., Bänziger, M., 1999. Selection improves drought tolerance in tropical maize populations: I. Gains in Biomass, Grain Yield, and Harvest Index. Crop Sci. 39:1306–1315.

Edmeades, G.O., McMaster, G.S., White, J.W., Campos, H. 2004. Genomics and the physiologist: bridging the gap between genes and crop response. Field Crops Res. 90:5-18.

Diker, K., Bausch, W.C., 2003. Potential use of nitrogen reflectance index to estimate plant parameters and yield of maize. Biosystems Eng. 85:437–447.

Drouet J.-L., Bonhomme, R., 1999. Do variations in local leaf irradiance explain changes to leaf nitrogen within row maize canopies? Ann. Bot.-London 84:61–69.

FAO. 2010. Food and Agriculture Organization of the United Nations. FAOSTAT. Available at http://faostat.fao.org/site/339/default.aspx. (verified 17 Feb. 2010).

Filella, I., Serrano, L., Serra, J., Penuelas J., 1995. Evaluating wheat nitrogen status with canopy reflectance indices and discriminant analysis. Crop Sci. 35:1400–1405.

Gastal F., Lemaire G., 2002. N uptake and distribution in crops: an agronomical and ecophysiological perspective. J. Exp. Bot. 53:789–799.

Gates, D., Keegan, H., Schleter, J., Weidner., V., 1965. Spectral properties of plants. Appl. Optics. 4:11–20.

Gitelson, A., Gritz, Y., Merzlyak., M., 2003. Relationships between leaf chlorophyll content and spectral reflectance and algorithms for non-destructive chlorophyll assessment in higher plant leaves. J. Plant Physiol. 160:271–282.

Graeff, S., Claupein, W., 2007. Identification and discrimination of water stress in wheat leaves (*Triticum aestivum* L.) by means of reflectance measurements. Irrigation Sci. 26:61–70.

Grant, L., 1987. Diffuse and specula characteristics of leaf reflectance. Remote Sens. Environ. 22:309-322.

Gutierrez, M., Reynolds, M.P., Raun, W. R., Stone, M.L., Klatt., A.R., 2010. Spectral water indices for assessing yield in elite bread wheat genotypes under well-irrigated, water-stressed, and high temperature conditions. Crop Sci. 50:197-214.

Hatfield, J.L., Gitelson, A.A., Schepers, J.S., Walthall, C.L., 2008. Application of spectral remote sensing for agronomic decisions. Agron. J. 100:117–131.

Hu, Y., Schmidhalter, U., 2005. Drought and salinity: A comparison of their effects on mineral nutrition of plants. J. Plant Nutr. Soil Sci. 168:541-549.

Huang, W., Wang, Z., Huang, L., Lamb, D.W., Ma, Z., Zhang, J., Wang, J., Zhao. C., 2010. Estimation of vertical distribution of chlorophyll concentration by bi-directional canopy reflectance spectra in winter wheat. Precision Agric DOI 10.1007/s11119-010-9166-5.

Hunt, E.R., Rock, B.N., 1989. Detection of changes in leaf water content using near- and middle-infrared reflectances. Remote Sens. Environ. 30:43–54.

Inoue, Y., Morinaga, S., Shibayama, M., 1993. Non-destructive estimation of water status of intact crop leaves based on spectral reflectance measurements. Jpn. J. Crop Sci. 62:462–469.

Keating B.A., Wafula, B.M., 1992. Modelling the fully expanded area of maize leaves. Field Crops Res. 29:163–176.

Lammel, J., Wollring, J., Reusch., S., 2001.Tractor based remote sensing for variable nitrogen fertilizer application. In W.J. Horst et al. (ed.) Plant nutrition – Food security and sustainability of agro-ecosystems. 694–695. Kluver Academic Publishers. Netherlands.

Liu, L., Zhao, C., Huang, W., Wang, J., 2003. Estimating winter wheat plant water content using red edge width. Int. J. Remote Sens. 25:3331–3342.

Major, D.J., Baumeister, R., Toure, A., Zhao, S., 2003. Methods of measuring and characterizing the effects of stresses on leaf and canopy signatures. ASA Spec. Publ. 66:165–175.

Mistele, B., Schmidhalter, U., 2008. Spectral measurements of the total aerial N and biomass dry weight in maize using a quadrilateral-view optic. Field Crops Res. 106:94–103.

Mistele, B., Schmidhalter, U., 2010. Tractor-based quadrilateral spectral reflectance measurements to detect biomass and total aerial nitrogen in winter wheat. Agron. J. 102:499–506.

Montes, J.M., Technow, F., Dhillon, B.S., Mauch, F., Melchinger, A.E., 2011. High throughput non-destructive biomass determination during early plant development in maize under field conditions. Field Crops Res. doi:10.1016/j.fcr.2010.12.017.

Neidhart, B., 1994. Morphological and physiological responses of tropical maize (*Zea mays* L.) to pre-anthesis drought. Ph.D. diss. Swiss Federal Institute of Technology Zürich, Switzerland.

Olfs, H.W., Blankenau, K., Brentrup, K., Jasper, J., Link, A., Lammel, J., 2005. Soil- and plant-based nitrogen-fertilizer recommendations in arable farming. J. Plant Nutr. Soil Sci. 168:414–431.

Osborne, S.L., Schepers, J.S., Francis, D.D., Schlemmer, M.R., 2002. Use of spectral radiance to estimate in-season biomass and grain yield in nitrogen- and water-stressed corn. Crop Sci. 42:165–171.

Penuelas, J., Gamon, J.A., Fredeen, A.L., Merino, J., Field, C.B., 1994. Reflectance indices associated with physiological changes in nitrogen- and water-limited sunflower leaves. Remote Sens. Environ. 48:135–146.

Penuelas, J., Pinol, J., Ogaya, R., Filella, I., 1997. Estimation of plant water concentration by the reflectance Water Index WI (R900/R970). Int. J. Remote Sens. 18:2869–875.

Peters, R.T., Evett, S.R., 2007. Spatial and temporal analysis of crop conditions using multiple canopy temperature maps created with center-pivot-mounted infrared thermometers. ASABE 50:919-927.

Poss, J.A., Russell, W.B., Grieve, C.M., 2006. Estimating yields of salt- and water-stressed forages with remote sensing in the visible and near infrared. J. Environ. Qual. 35:1060–1071.

Reusch, S., Link, A., Lammel., J., 2002. Tractor-mounted multispectral scanner for remote field investigation. In P.C. Roberts (ed.) Proceedings of the 6th International Conference on Precision Agriculture, Minneapolis. ASA-CSSA-SSSA, Madison, WI, USA, pp. 1385–1393.

Reusch, S., 2003. Optimisation of oblique-view remote measurement of crop N-uptake under changing irradiance conditions. Precision Agriculture. Hydro Agri, Research Centre Hanninghof, Dülmen, Germany.

Reynolds, M., Manes, Y., Izanloo, A., Langridge, P., 2009. Phenotyping approaches for physiological breeding and gene discovery in wheat. Ann. Appl. Biol. 155:309–320.

Richards, R.A., Rebetzke, G.J., Watt, M., Condon, A.G., Spielmeyer, W., Dolferus, R., 2010. Breeding for improved water productivity in temperate cereals: Phenotyping, quantitative trait loci, markers and the selection environment. Funct. Plant Biol. 37:85–97.

Rollin, E.M., Milton, E.J., 1998. Processing of high spectral resolution reflectance data for the retrieval of canopy water content information. Remote Sens. Environ. 65:86–92.

Samborski S.M., Tremblay N., Fallon E., 2009. Strategies to make use of plant sensors-based diagnostic information for nitrogen recommendations. Agron J 101:800–816

Schlemmer, M.R., Francis, D.D., Shanahan, J.F., Schepers, J.S., 2005. Remotely measuring chlorophyll content in corn leaves with differing nitrogen levels and relative water content. Agron. J. 97:106–112.

Schmidhalter, U., Glas, J., Heigl, R., Manhart, R., Wiesent, S., Gutser, R., Neudecker. E., 2001. Application and testing of a crop scanning instrument – field experiments with reduced crop width, tall maize plants and monitoring of cereal yield. In G. Grenier et al. (ed.) Proceedings of the 3rd European Conference on Precision Agriculture. Montpellier, France. pp. 953-958.

Schmidhalter, U., Jungert, S., Bredemeier, C., Gutser, R., Manhart, R., Mistele, B., Gerl, G., 2003. Field-scale validation of a tractor based multispectral crop scanner to determine biomass and nitrogen uptake of winter wheat. In: Stafford, J., Werner, A. (Eds.), Precision Agriculture: Proceedings of the 4th European Conference on Precision Agriculture. Academic Publishers, Wageningen, pp. 615–619.

Schmidhalter, U., 2005. Sensing soil and plant properties by non-destructive measurements. Proceedings of the International Conference on Maize Adaption to Marginal Environments. 25th Anniversary of the Cooperation between Kasetsart University and Swiss Federal Institute of Technology. March 6-9. Nakhon Ratchasima, Thailand.

Schmidhalter, U., Bredemeier, C., Geesing, D., Mistele, B., Selige, T., Jungert, S., 2006. Precision agriculture: spatial and temporal variability of soil water, soil nitrogen and plant crop response. Bibliotheca Fragmenta Agronomica 11:97–106.

Schmidt, W., 1995. Competition effects in silage maize surveys. (In German). Ph.D. diss. Technische Universität München, Germany.

Seelig, H.D., Adams III, W.W., Hoehn, A., Stodieck, L.S., Klaus, D.M., Emery, W.J., 2008a. Extraneous variables and their influence on reflectance-based measurements of leaf water content. Irrigation Sci. 26:407–414.

Seelig, H.D., Hoehn, A., Stodieck, L.S., Klaus, D.M., Adams III, W.W., Emery, W.J., 2008b. Relations of remote sensing leaf water indices to leaf water thickness in cowpea, bean, and sugarbeet plants. Remote Sens. Environ. 112:445–455.

Seelig, H.D., Hoehn, A., Stodieck, L.S., Klaus, D.M., Adams III, W.W., Emery, W.J., 2009. Plant water parameters and the remote sensing R1300/R1450 leaf water index: controlled condition dynamics during the development of water deficit stress. Irrigation Sci. 27:357–365.

Serraj, R., McNally, K. L., Slamet-Loedin, I., Kohli, A., Haefele, S. M., Atlin, G., Kumar, A., 2011. Drought resistance improvement in rice: an integrated genetic and resource management strategy. Plant Prod. Sci. 14(1): 1–14.

Soler, C.M.T., Hoogenboom, G., Sentelhas, P.C., Duarte, A.P., 2007. Impact of water stress on maize grown off-season in a subtropical environment. J. Agron. Crop Sci. 193:247—261.

Thoren, D., Schmidhalter, U., 2009. Nitrogen status and biomass determination of oilseed rape by laser-induced chlorophyll fluorescence. Eur. J. Agron. 30:238–242.

Tilling, A.K., O'Leary, G.J., Ferwerda, J.G., Jones, S.D., Fitzgerald, G.J., Rodriguez, D., Belford, R., 2007. Remote sensing of nitrogen and water stress in wheat. Field Crop Res. 104:77–85.

Valentinuz, O.R., Tollenaar, M., 2006. Effect of genotype, nitrogen, plant density, and row spacing on the area-per-leaf profile in maize. Agron. J. 98:94–99.

Wang, Z., Wang, J., Zhao, C., Zhao, M., Huang, W., Wang, C., 2005. Vertical distribution of nitrogen in different layers of leaf and stem and their relationship with grain quality of winter wheat. J. Plant Nutr. 28:73–91.

Winterhalter, L., Mistele, B., Jampatong, S., Schmidhalter, U., 2011. High throughput sensing of aerial biomass and above ground nitrogen uptake in the vegetative stage of well-watered and drought stressed tropical maize hybrids. Crop Sci. 51:1–11.

Worku, M., Bänziger, M., Schulte auf 'm Erley, G., Friesen, D., Diallo, A.O., Horst, W.J., 2007. Nitrogen uptake and utilization in contrasting nitrogen efficient tropical maize hybrids. Crop Sci. 47:519–528.

Yu, G., Miwa, T., Nakayama, K., Matsuoka, N., Kon, H., 2000. A proposal for universal formulas for estimating leaf water status of herbaceous and woody plants based on spectral reflectance properties. Plant Soil 227:47–58.

List of Publications

Publications

I. **Winterhalter, L.**, Mistele, B., Jampatong, S., Schmidhalter, U. 2011. High throughput sensing of aerial biomass and above ground nitrogen uptake in the vegetative stage of well-watered and drought stressed tropical maize hybrids. Crop Science 51, 479-489.

II. **Winterhalter, L.**, Mistele, B., Jampatong, S., Schmidhalter, U. 2011. High throughput phenotyping of canopy water mass and canopy temperature in well-watered and drought stressed tropical maize hybrids in the vegetative stage. European Journal of Agronomy 35, 22-32.

III. **Winterhalter, L.**, Mistele, B., Schmidhalter, U. 2011. Assessing the vertical footprint of reflectance measurements to characterize nitrogen uptake and biomass distribution in maize canopies. Field Crops Research 129, 14-20.

Appendix

Publication I

High Throughput Sensing of Aerial Biomass and Above Ground Nitrogen Uptake in the Vegetative Stage of Well-Watered and Drought Stressed Tropical Maize Hybrids

ABSTRACT

The measurement of agronomical parameters of maize (*Zea mays* L.) indicating its biomass and nutritional status provides important information to understand its responses to the environment. The detection of significant differences among maize hybrids would be very useful in plant breeding programs screening for nitrogen uptake and drought tolerance. The aim of the study was to assess the efficacy of high throughput sensing measurements to determine the aerial biomass and nitrogen uptake of tropical maize hybrids grown in well-watered (control) and drought stress treatments. Experiments were conducted at the National Corn and Sorghum Research Center in Thailand in the years 2007-2009. High throughput canopy reflectance measurements using spectral indices from the literature and newly developed for this study were performed regularly along with biomass samplings until flowering. The relationship of the spectral indices with each of aerial biomass and nitrogen uptake had coefficients of determination of up to 0.8 and were also able to distinguish between drought stress levels. Through most sampling dates and stress levels, varieties were similarly classified in their amount of aerial biomass and nitrogen uptake by destructive and non-contacting measurements. Our results support the possibility of incorporating these methods in the development of high throughput phenotyping techniques that could prove to be potentially useful for future plant breeding.

Global concern exists regarding our capacity to feed a fast increasing world population against a background of climate change and shortage of water for agriculture (Chaves and Davies, 2010). The effective use of limited water resources by crops is the basis of drought resistance. This can be achieved by either adjusting crop phenology to its environment or by using agronomic practices aimed at improved water use (Passioura, 2007). Among the major agricultural crops, the worldwide annual harvest quantities of maize lie clearly before those of rice or wheat (784, 651, and 607 million tons in 2007, respectively; FAO, 2010). As a consequence, the continued successful cultivation of maize is crucial for global food security, although significant yield losses from drought are expected to increase because of global climate change (Campos et al., 2004).

The measurement of agronomical traits of maize to indicate its biomass and nutritional status provides important information to understand its response to the environment. For example, both N deficiencies and water deficiencies negatively affect the amount of chlorophyll produced in the plants as well as cell turgidity. As the water deficit increases, plants appear increasingly wilted and the photochemical activity of chlorophyll is reduced, while water stress also has a nonlinear effect on the growth and development of plants (Clay et al., 2006; Yu et al., 2000). The interaction of water and N stress has a strong influence on crop yield (Hu and Schmidhalter, 2005), with N uptake being dependent on fertility of soil and plant water status.

Remote sensing represents a valuable tool with the potential to assess rapidly a wide variety of physiological properties (Schlemmer et al., 2005), whereas spectral reflectance indices could be used as indicators of plant status under different stresses (Penuelas et al., 1994). Furthermore, the application of this technique would be invaluable in plant breeding for screening tests of drought tolerant plants (Inoue et al., 1993; Schmidhalter, 2005).

Previous studies indicate that the strongest relationships between leaf reflectance and chlorophyll content occur in the either the green spectrum near 550 mm or the far red spectrum near 700 nm (Gitelson et al. 2003; Mistele and Schmidhalter, 2010b). To estimate aerial biomass using spectral techniques, measurements in the near infrared (NIR) are most useful. Reflectance in this range is mainly characterized by the difference in the optical densities between the saturated cell walls and the intercellular region (Gates et al., 1965). Drought stress changes the leaf cell structure and composition, primarily the properties of the connections between cell walls and air spaces, cell sizes and shapes, cell wall composition and structure (Grant, 1987; Liu et al., 2003; Penuelas et al., 1994). Thus, the plant water status directly influences the intercellular air spaces, as well as the cell turgidity and consequently the leaf cell structure that affects how light will be absorbed, transmitted, or reflected by leaves (Schlemmer et al., 2005; Seelig et. al, 2008b), leading to variation in leaf spectral reflectance. Specifically, a reduction in the amount of water in the leaves tends to increase the reflectance in the 400 - 1300 nm region (Graeff and Claupein, 2007; Inoue et al., 1993).

Moreover, the quantity of light absorbed by leaves is a function of its photosynthetic pigment content. Therefore, the chlorophyll content not only directly determines photosynthetic potential and primary production, but also provides an indirect estimate of the nutrient status because the leaf enzyme Rubisco as the main sink for N is closely tied to chlorophyll (Chapelle et al., 1984), thus affecting the leaf absorption spectra. With increasing chlorophyll content, the visible wavelength absorption increases, whereas leaf reflectance decreases (Hatfield et al., 2008). Importantly, the reflectance in the red edge region contains information about chlorophyll absorption and cell wall reflection, plus the alteration between these main effects. With increasing N content, the reflection as well as the inflection point in the red edge changes (Mistele and Schmidhalter, 2008).

Based on these observations, vegetation indices have been developed to relate reflectance of plant canopies with canopy characteristics to estimate agriculturally important plant parameters. Among the easiest and most commonly used spectral indices are two-wavelength ratio vegetation indices (Diker and Bausch, 2003; Gutierrez et al., 2010; Reusch, 2003). Use of these and other similar indices can give good predictions of total aerial dry matter and aboveground N-uptake (Mistele and Schmidhalter, 2010b), both of which are useful indicators of plant performance playing an important role in plant growth. Spectral indices derived from real-time multispectral reflectance information also have the potential to be used as indicators of plant water status and deficit stress (Claudio et al., 2006; Seelig et al., 2009).

Our capability and capacity to measure plant phenotypes for important traits has not kept pace with the increasing number of genotyping initiatives in plant species over the past decade, with this delay hindering our ability to describe gene-to-phenotype relationships for many important, but complex traits in various crop plants. A cogent example is drought tolerance in maize, where the major limitation to improved selection methods for water-limited environments is the lack of fast and precise measurements of the plant phenotype. Thus, high throughput precision phenotyping would provide researchers with the means to gain detailed, reliable information of plant characteristics describing the trait phenotypes for the many genotypes that a typical plant breeding population contains (Campos et al., 2004; Thoren and Schmidhalter, 2009). More importantly, its combination with high throughput genotyping promises to be a major breakthrough in our goal of explaining the underlying genetic basis of complex plant traits such as drought tolerance (Montes et al., 2007). It should thus be clear that improvements in the speed and precision of trait phenotyping as well as a direct selection for the phenotype will have a greater impact on crop improvement than further advances in molecular technologies (Richards et al., 2010).

In meeting this goal, however, it is important to realize that the relationship of spectral reflectance and N status differs between crop varieties. To compensate for this effect, reference values are necessary to determine the fertilizer application rate (Lammel et al., 2001), which, in turn, could be used by breeders to differentiate between varieties. Indeed, in light of the evaluation of large populations in plant breeding having to be done as quickly and as cheaply as possible, the early selection in reducing the size of populations has to be reliable and effective (Richards et al., 2010). The major challenge for remote sensing researchers is to fully understand its potential in applications such as these. Solving these problems requires an extension of the background knowledge of the agronomic information content of remote sensing data, as well as algorithmic innovations to interpret these data (Hatfield et al., 2008).

The aim of this study was to employ high throughput sensing measurements (and a selection of spectral indices) to determine the aerial biomass and N uptake of several tropical maize hybrids experiencing different drought stress treatments. This was done with a view towards developing a technique that could prove to be useful for future breeding.

MATERIALS AND METHODS

Experimental design

The experiment was conducted at the National Corn and Sorghum Research Center (101.3 E longitude; 14.6 N latitude; 380 m altitude), located 155 km northeast of Bangkok in the tropical region of Thailand. The climate is influenced by annual rainy and dry seasons, the latter generally lacking any rainfall. Trials were carried out in the dry seasons (November to February) in the years 2007/08 and 2008/09. The soil of the research station is classified as reddish brown lateritic soil with moderately high water permeability (Neidhart, 1994).

Seven maize hybrids, four irrigation treatments and four replications, giving a total of 112 plots were arranged as a randomized block design. Each plot consisted of 10 rows, with a distance of 0.20 m between the plants and 0.75 m between the rows and a plot length of 10 m. The seven hand-sown tropical high yield hybrids were selected from different sources (Table 1). Different drought stress levels were represented by each of the four irrigation treatments, consisting of a control treatment with weekly irrigation and three stress treatments that withheld irrigation for various periods (Table 2). In all treatments, sprinkler irrigation was applied until 3 weeks after sowing, followed by furrow irrigation. Fertilization was applied as basal fertilizer with two components (N+P) at a rate of 20 kg N ha^{-1} and urea as side-dress at a rate of 115 kg N ha^{-1}, leading to a total fertilization rate of 135 kg N ha^{-1}. Cultivation methods (e.g., thinning and hand weeding) were done following local technical recommendations.

Table 1. Tropical maize hybrids used in this study.

No.	Hybrid	Source
1	Pac 224	Pacific Seeds (Thai) Limited, Saraburi, Thailand
2	NK 40	Syngenta Seeds Limited, Bangkok, Thailand
3	30Y87	Pioneer Hi-Bred (Thailand) Co., Ltd., Bangkok, Thailand
4	CP AAA Super	Bangkok Seeds Industry Co., Ltd., Bangkok, Thailand
5	DK 979	Monsanto Seeds (Thailand) Limited, Phitsanulok, Thailand
6	NS 2	Nakhon Sawan Field Crops Research Center, Nakhon Sawan, Thailand
7	Suwan 4452	National Corn and Sorghum Research Center, Pakchong, Thailand

Table 2. Irrigation treatments and important dates in the experimental years 2007/08 and 2008/09.

Irrigation treatment	Experiment 2007/08	Experiment 2008/09
Control (C)	weekly irrigation	weekly irrigation
early vegetative stress (ES)	no irrigation: 07 Dec. 2007 - 03 Jan. 2008	no irrigation: 13 Dec. 2008 - 08 Jan. 2009
late vegetative stress (LS)	no irrigation: 22 Dec. 2007 - 17 Jan. 2008	no irrigation: 20 Dec. 2008 - 15 Jan. 2009
pre-flowering stress (PS)	no irrigation: 29 Dec. 2007 - 24 Jan. 2008	no irrigation: 27 Dec. 2008 - 29 Jan. 2009
Important dates		
sowing date	13 Nov. 2007	18 Nov. 2008
onset of flowering	18 Jan. 2008	29 Jan. 2009

Plant and soil water status measurements

Biomass harvests were performed regularly until flowering to determine aerial biomass and N uptake. At each harvest, 30 plants were cut from 6 m in the inner rows (not border plants) and their fresh weight was determined. The plant material was then chopped, with a representative subsample being collected and weighed before being oven dried at 100 °C for 3 days and reweighed. The dried samples were ball milled (100 µm) and analyzed for total N content with an Isotope Radio Mass Spectrometer (IRMS) combined with a preparation unit (ANCA SL 20-20, Europe Scientific, Crewe, UK).

Tensiometers were used to monitor the soil matric potential and to control the drying out of the soil during the drought stress periods. The tensiometers were installed at depths of 20, 40, 60, 80 and 100 cm with the measurements being taken regularly until flowering. All measurements were replicated three times as a double set in a subset of two hybrids in each of the control, early vegetative stress and late vegetative stress treatments.

Spectral measurements

Canopy reflectance measurements were coupled with the biomass samplings, always being taken at the same time (between 11:00 and 13:00). The sensor system was mounted on a forklift, enabling us to vary the height of the sensor to be able to measure fully developed maize canopies with plant heights of up to 2.5 m. The sensor contains two diode-array spectrometer units that measure reflectance and incident radiation simultaneously. One component was connected to a four-in-one light fiber to create an optical mixed signal of the canopy reflection of one area as taken from four directions with an oblique quadrilateral-view, whereas the second component was connected to a diffuser to measure the global radiation to account for different light conditions. The field of view included three to five maize rows depending on plant height. Of the two spectrometers, one analyzed the reflected radiation in 256 spectral channels with a detection range from 300 to 1000 nm and a bandwidth of 3.3 nm, whereas the second had a detection range from 1000 to 1700 nm and a bandwidth of 6 nm (Mistele and Schmidhalter, 2008).

Spectral indices

Spectral indices, often calculated with the reflectance values from two wavelengths, have been used to describe the relationship of agronomical traits of plant canopies (e.g., plant biomass, N and chlorophyll) with their reflectance (Mistele and Schmidhalter, 2008; Rodriguez et al., 2006; Tilling et al., 2007). To evaluate the sensor system, a wide range of spectral indices, taken from both the literature as well as being newly developed, were related to the measured aerial biomass and N uptake. A detailed overview of the spectral indices is given in Table 3.

Table 3. Spectral reflectance indices evaluated in this study. The parameters of primary sensitivity are: DM = dry matter yield (kg/ha); N = total aerial nitrogen (kg/ha); %N = nitrogen content (%); Chl = chlorophyll content (mol/m^2); %Chl = Chlorophyll concentration (%); LAI = leaf area index; PWC = plant water content (%); RWC = relative water content (%); EWT = equivalent water thickness (μm); and YLWS & YLNS = yield losses (kg/ha) due to water and N stress, respectively.

Index	Formula	Parameter	Plant species	Plant level	Growth condition	Author
760/730	R_{760}/R_{730}	N, DM	maize	canopy	field	Mistele and Schmidhalter (2010a)
SR (simple ratio)	R_{900}/R_{680}	LAI	durum wheat	canopy	field	Aparicio et al. (2002); Poss et al. (2006)
WBI/NDVI	$[R_{900}/R_{970}]/[(R_{800}-R_{680})/(R_{800}+R_{680})]$	PWC	trees, shrubs, grass	canopy	field	Claudio et al. (2006); Penuelas et al. (1997)
WBI (Water Band Index)	R_{900}/R_{970}	PWC	trees, shrubs, grass	canopy	field	Claudio et al. (2006); Penuelas et al. (1997)
NDWI (Normalized Difference Water Index)	$(R_{840}-R_{1650})/(R_{840}+R_{1650})$	YLWS & YLNS	maize	canopy	field	Clay et al. (2006)
NIR/NIR	R_{780}/R_{740}	N, DM	maize	canopy	field	Mistele and Schmidhalter (2008)
NIR/red	R_{780}/R_{700}	N, DM	maize	canopy	field	Mistele and Schmidhalter (2008)
NIR/green	R_{780}/R_{550}	N, DM	maize	canopy	field	Mistele and Schmidhalter (2008)
NDVI (Normalized Difference Vegetation Index)	$(R_{800}-R_{680})/(R_{800}+R_{680})$	N, DM, %N, %Chl	maize, wheat	canopy	field	Mistele and Schmidhalter (2008); Tilling et al. (2007)
440/685	R_{440}/R_{685}	RWC, Chl	maize	leaf	pot	Schlemmer et al. (2005)
525/685	R_{525}/R_{685}	RWC, Chl	maize	leaf	pot	Schlemmer et al. (2005)
YCAR	R_{600}/R_{680}	RWC, Chl	maize	leaf	pot	Schlemmer et al. (2005)
OCAR	R_{630}/R_{680}	RWC, Chl	maize	leaf	pot	Schlemmer et al. (2005)
1300/1450	R_{1300}/R_{1450}	EWT	tree, bean	leaf	field, pot	Seelig et al. (2008a)
NDRE (Normalized Difference Red Edge)	$(R_{790}-R_{720})/(R_{790}+R_{720})$	%N, %Chl	wheat	canopy	field	Rodriguez et al. (2006); Tilling et al. (2007)

Statistical Analysis

Statistical analysis was conducted to determine the effect of hybrid and different irrigation treatments on the relationship between the spectral indices and either of aerial biomass or N uptake. Curvilinear models (quadratic) in Microsoft Excel 2003 (Microsoft Inc., Seattle, WA, USA) as well as analysis of variance (ANOVA) and general linear model (GLM) analysis in SPSS 16 (SPSS Inc., Chicago, USA) were used to establish relationships.

RESULTS

Soil water status

Changes in the soil matric potential in the experimental year 2007/08 for the different irrigation treatments are shown in Fig. 1; comparable results were obtained in the experimental year 2008/09 (data not presented). In the control treatment (Fig. 1a), the weekly irrigation events meant that only a slight drying of the upper 20 cm of the soil was observed. By contrast, a strong drying out was observed in the upper 20 cm of the soil for the early vegetative stress treatment (Fig. 1b), together with a slight drying in the 21- to 40-cm layer. This trend was expanded in the late vegetative stress treatment (Fig. 1c), which showed a strong drying out of the soil down to a depth of 60 cm and a slight drying in the 61- to 100-cm layer.

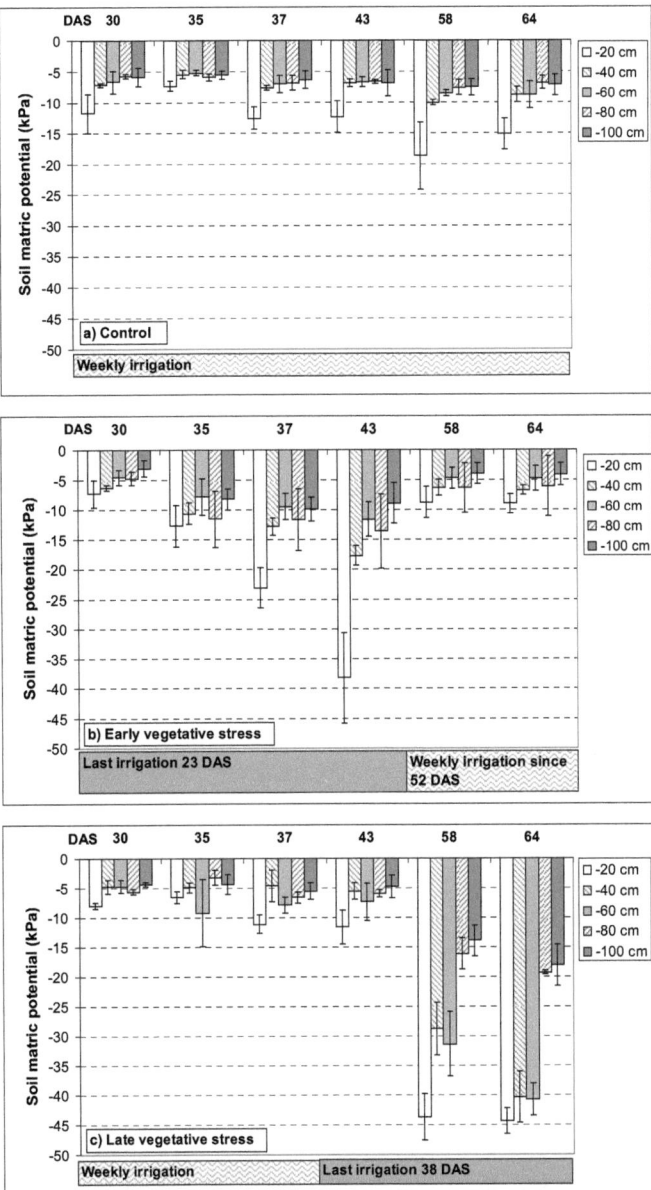

Figure 1. Soil matric potential measured at different depths using tensiometers for three irrigation treatments in the experimental year 2007/08: control (a), early vegetative stress (b) and late vegetative stress (c). Error bars indicate standard deviations.

Influence of maize hybrids on aerial biomass and N uptake

A full ANOVA analysis indicated that all tested effects were highly significant, with highest F-values being observed for sampling date, followed by year, irrigation treatments and cultivar effects. Interaction effects were also highly significant except for cultivar*irrigation treatment, cultivar*irrigation treatment*year and cultivar*irrigation treatment*sampling date*year. In subsequent analyses years and sampling dates were evaluated separately. Years were evaluated separately because of the cooler weather in the second experimental year, leading to smaller plants during the vegetation period. The purpose of the analysis was to show possible differences among hybrids at specific sampling dates, hence they were analyzed individually.

Irrigation treatments and tropical maize hybrids had individually a significant influence on the aerial biomass and N uptake for all measurements in the GLM analysis for the experimental years 2007/08 and 2008/09, except for the cultivars 56 DAS (2008/09) and irrigation treatments 35 DAS (Table 4). There was no influence on the interaction of hybrids with irrigation treatments.

For all sampling dates and irrigation treatments, varieties were similarly classified in their amount of aerial biomass and N uptake as exemplified in Fig. 2 and 3 for the biomass harvest 56 DAS in the experimental year 2007/08. A statistically based grouping of the hybrids is presented in Table 5. Significant differences in aerial biomass and N uptake were found only for hybrids with the maximum and minimum values in the experimental years 2007/08 and 2008/09. The hybrids 6 and 7 had significantly higher aerial biomass, N uptake and index values than the hybrids 2 and 5 in 2007/08, while in 2008/09 a differentiation was possible only for the first two sampling dates.

Table 4. Results from an analysis of variance of the influence of hybrid and irrigation treatment on values of each of aerial biomass and N uptake in the experimental years 2007/08 and 2008/09.

Days after sowing	Parameter	Aerial biomass		N uptake	
		df	F value	df	F value
Experiment 2007/08					
35	Hybrid	6	11***	6	10.8***
	Irrigation treatment	1	1.8	1	7.3**
	Hybrid × Irrigation treatment	12	0.2	12	0.2
42	Hybrid	6	14.3***	6	11.2***
	Irrigation treatment	2	18***	2	38***
	Hybrid × Irrigation treatment	12	0.8	12	0.8
56	Hybrid	6	13.1***	6	6.2***
	Irrigation treatment	3	49.3***	3	35***
	Hybrid × Irrigation treatment	18	0.4	18	0.5
63	Hybrid	6	16.4***	6	12.3***
	Irrigation treatment	3	87.9***	3	73.3***
	Hybrid × Irrigation treatment	18	0.9	18	0.5
Experiment 2008/09					
35	Hybrid	6	5.1***	6	4.2***
	Irrigation treatment	2	0.3	2	0.1
	Hybrid × Irrigation treatment	12	0.6	12	0.6
49	Hybrid	6	6.5***	6	5.3***
	Irrigation treatment	3	3.9**	3	5.7***
	Hybrid × Irrigation treatment	18	0.5	18	0.5
56	Hybrid	6	1.9	6	1.3
	Irrigation treatment	3	8.8***	3	9.5***
	Hybrid × Irrigation treatment	18	0.3	18	0.3
63	Hybrid	6	2.6*	6	2.2*
	Irrigation treatment	3	4**	3	6.2***
	Hybrid × Irrigation treatment	18	0.9	18	0.5

*Significant at the 0.05 level.

**Significant at the 0.01 level.

***Significant at the 0.001 level.

Table 5. Results from an analysis of variance of aerial biomass, N uptake and index NIR/NIR (R_{780}/R_{740}) of seven maize hybrids (average of all irrigation treatments) for the experimental years 2007/08 and 2008/09. Letters (a, b, c) indicate groups of hybrids that were significantly different from each other at the 0.05 level (SNK test).

Hybrid No.	Year 2007/08					Hybrid No.	Year 2008/09				
	All harvests	DAS					All harvests	DAS			
	Aerial biomass	35	42	56	63		Aerial biomass	35	49	56	63
7	a	a	a	a	a	7	a	a	a	a	a
6	a	a	a	a	a	6	a	a, b	a	a	a
1	a	b	b	a, b	a, b	2	a	b	a, b	a	a
4	a	b, c	b	a, b	a, b	4	a	b	a, b	a	a
3	a	b, c	b	b	a, b	1	a	b	b, c	a	a
5	a	b, c	b	b	b	3	a	b	b, c	a	a
2	a	c	b	b	b	5	a	b	c	a	a
	N uptake	35	42	56	63		N uptake	35	49	56	63
6	a	a	a	a	a	7	a	a	a	a	a
7	a, b	a	a	a, b	a	6	a	a, b	a, b	a	a
1	a, b, c	b	b	a, b	a, b	2	a	a, b	a, b	a	a
3	a, b, c	b	b	a, b	a, b	4	a	b	a, b	a	a
4	a, b, c	b	b	a, b	a, b	1	a, b	b	b, c	a	a
2	b, c	b	b	b	b	3	a, b	b	b, c	a	a
5	c	b	b	b	b	5	b	b	c	a	a
	R_{780}/R_{740}	36	43	58	64		R_{780}/R_{740}	37	50	58	64
6	a	a	a	a	a	6	a	a, b	a	a	a
7	b	a, b	a, b	a, b	a, b	4	a	b, c	a	a	a
3	b, c	b, c	b, c	b	a, b	2	a	a, b	a	a	a
1	b, c	b, c	b, c	b	a, b	7	a	a	a	a	a
4	b, c	b, c	b, c	b	a, b	1	a, b	c	a, b	a	a
2	b, c	c	c	b	b	3	a, b	c	a, b	a	a
5	c	c	c	b	b	5	b	c	b	a	a

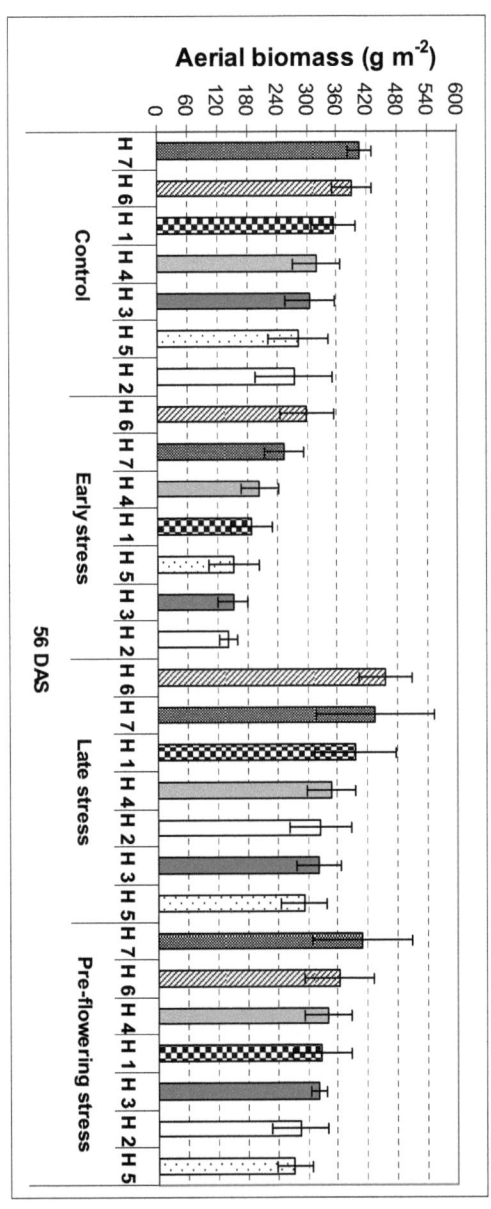

Figure 2. Aerial biomass of seven maize hybrids at different irrigation treatments 56 days after sowing for the experimental year 2007/08. Error bars indicate standard deviations.

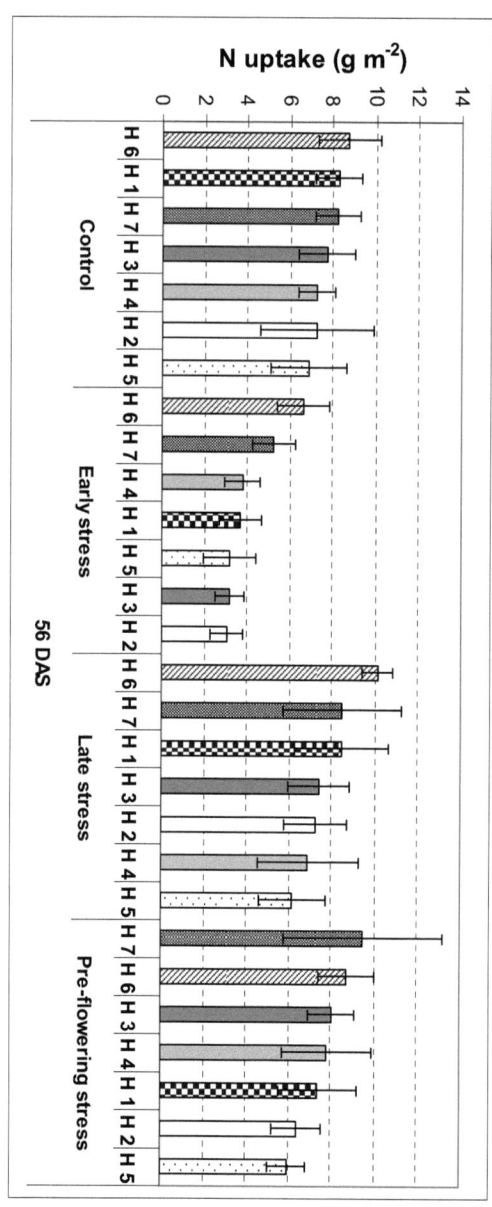

Figure 3. Nitrogen uptake of seven maize hybrids at different irrigation treatments 56 days after sowing for the experimental year 2007/08. Error bars indicate standard deviations.

Irrigation treatment and hybrid effects on spectral indices

An overview of the quality for the relationship of aerial biomass, N uptake and the best spectral indices from literature as well as the newly developed for this study is given in Fig. 4 for both experimental years. Our goal here was to find those spectral indices that could detect differences among the irrigation treatments given that this variable was found to have a significant influence on the aerial biomass and N uptake (see above). Lower R^2 values were found on average for the NDVI (Normalized Difference Vegetation Index) with 0.56 for aerial biomass and 0.51 for N uptake, while the values for the other indices not presented in Fig. 4 were between 0.42 – 0.57 for aerial biomass and 0.37 – 0.55 for N uptake. The relationship between aerial biomass and the spectral index Normalized Difference Red Edge (NDRE) at 57 DAS in the experimental year 2007/08 (Fig. 5a), as well as that between N uptake and the spectral index R_{760}/R_{730}, is shown in Fig. 5b to exemplify the results. The coefficients of determination in both cases were 0.74 over all treatment groups, with lower R^2 values for almost all individual irrigation treatments. The exception here was for the early vegetative stress treatment (N uptake) with an R^2 of 0.79. The more water that was available for the plants through irrigation, the higher the index value; the index values also increased for all measurements with increasing aerial biomass and N uptake values.

The relationship of aerial biomass and the spectral index R_{780}/R_{740} for the seven hybrids showed R^2 values of up to 0.88, indicating the close relationship of index values and aerial biomass for each hybrid (e.g., see Fig. 5c). Similar, strong results were obtained for N uptake (Fig. 5d), with R^2 values of up to 0.86. With increasing aerial biomass and N uptake, the spectral index values again increased for all measurements.

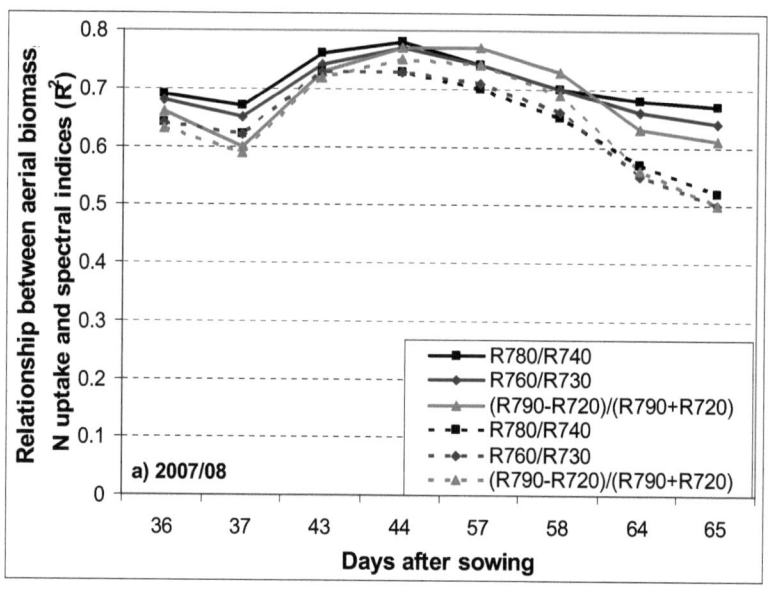
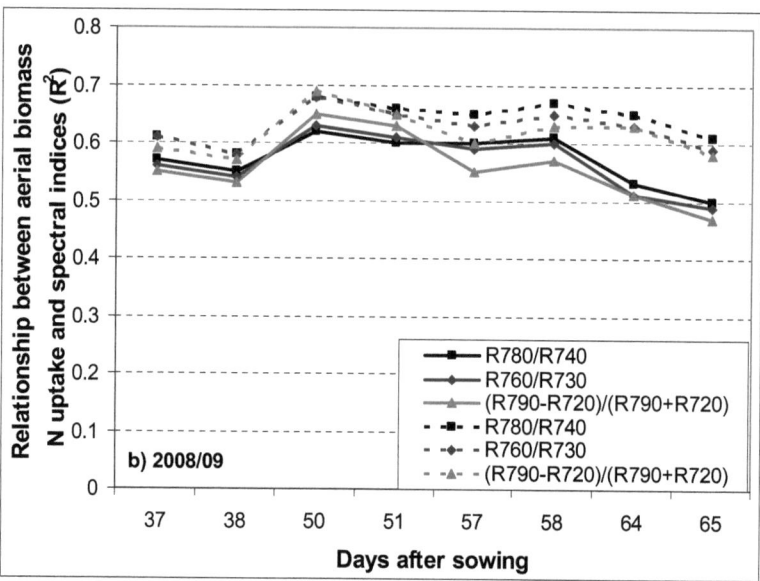

Figure 4. Coefficients of determination (R^2) between spectral indices and each of aerial biomass (dotted lines) and N uptake (continuous lines) for seven tropical maize hybrids and four irrigation treatments for the experimental years 2007/08 (a) and 2008/09 (b).

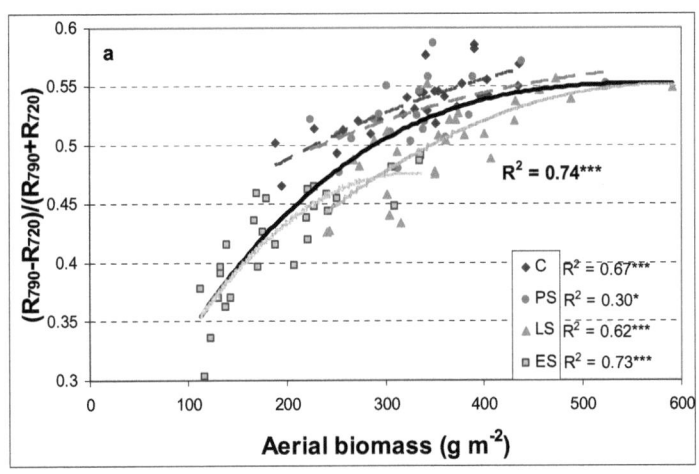
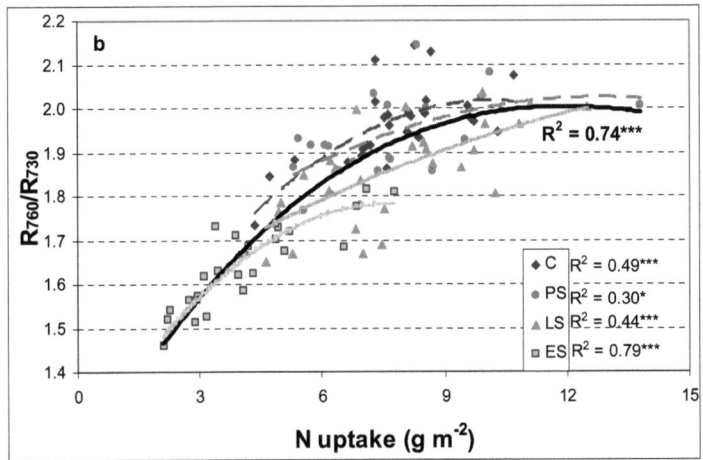

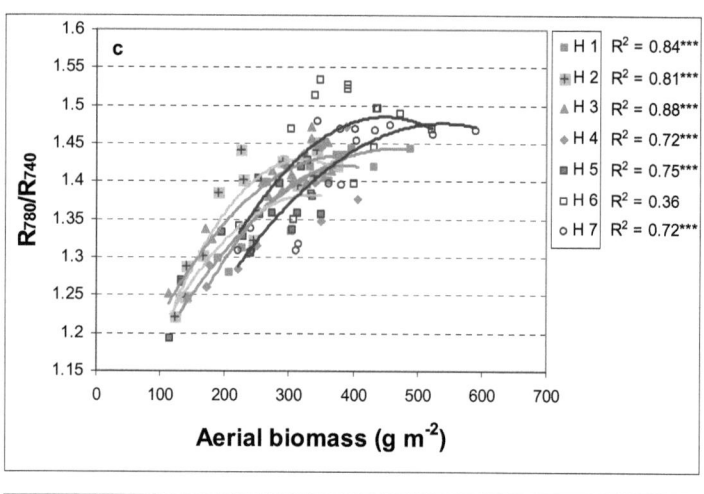

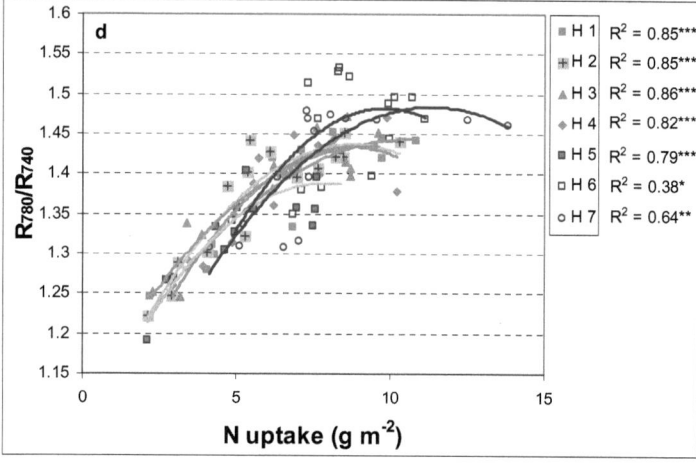

Figure 5. Relationship of aerial biomass and the spectral index NDRE [(R790-R720)/(R790+R720)] (a), as well as N uptake and the spectral index R_{760}/R_{730} (b) at 57 days after sowing in the experimental year 2007/08. Coefficients of determination are presented averaged over all hybrids and irrigation treatments (0.74) as well as for the different irrigation treatments (C = control, PS = pre-flowering stress, LS = late vegetative stress and ES = early vegetative stress). Relationship of aerial biomass and spectral index NIR/NIR (R_{780}/R_{740}) (c), as well as N uptake and spectral index NIR/NIR (R_{780}/R_{740}) (d) at 57 days after sowing in the experimental year 2007/08. Coefficients of determination are presented individually for the seven hybrids. *Significant at the 0.05 level, **Significant at the 0.01 level, ***Significant at the 0.001 level.

DISCUSSION

Soil water status

Results of the tensiometer measurements illustrated the changes in soil matric potential in the experimental treatments, showing the drying out of the soil with increasing drought stress and indicating the major rooting depth and water uptake of the maize hybrids. Our observations agree with the results of Soler et al. (2007), who reported that tensiometers installed in drought stressed maize stands at a soil depth of 20 cm were the first ones to show a decrease in the soil matric potential, followed by decreases at soil depths of 40 cm around 30 days after sowing, indicating the rapid, initial consumption of water in uppermost soil layers. Measurements of soil water status were very important here to back up sensor measurements of the drought stressed hybrids. The experimental site was characterized by a low available soil water capacity demonstrating only moderate decreases in soil matric potential even at severe drought stress (Camp, 1996).

Differentiation and detection of aerial biomass and nitrogen uptake of well-watered and drought stressed tropical hybrids

Previous studies have indicated that the spectral reflectance of maize canopies is influenced by chlorophyll content and N status of the plant (Hatfield et al., 2008; Mistele und Schmidhalter, 2008; Schlemmer et al., 2005); drought and N stress also impact the reflectance of plants at different growth stages (Clay et al., 2006). Therefore, indices that detect the pigment composition reflecting leaf nutrient and water status could assess physiological changes associated with these stresses (Penuelas et al., 1994) by measuring changes in crop reflectance of plants in specific wavelength ranges such as the visible wavelength range (Hatfield et al., 2008), the red edge, and the NIR band (Liu et al., 2003; Yu et al., 2000). The results in this study showed that the spectral indices were successful at detecting the significant influence of the different irrigation treatments on the aerial biomass and N uptake of seven different tropical maize hybrids. Several indices (R_{780}/R_{740}; ($R_{790}-R_{720}$)/($R_{790}+R_{720}$); R_{760}/R_{730}) showed particularly strong relationships (R^2 values of up to 0.8) during the entire vegetative stage of tropically grown maize. Similar results were reported by Mistele and Schmidhalter (2008) for temperate maize. The relationship of the index NIR/NIR (R_{780}/R_{740}) with total aerial N had average R^2 values of 0.79. Our results are in agreement with those from Diker and Bausch (2003), who showed that the vegetation indices correlate well with dry matter at different growth stages, albeit with variable performance in estimating the dry matter production of maize, which seems to be growth-stage dependent. In addition, it is worth noting that the experimental year also had an influence on the quality of indices to assess total aerial N and dry matter yield.

The relationships of aerial biomass and the spectral index $(R_{790}-R_{720})/(R_{790}+R_{720})$ and of N uptake and the spectral index R_{760}/R_{730} both had high coefficients of determination (0.74) across all drought stress treatments, and also showed clear differences between the treatments, demonstrating the ability of the indices to accurately assess aerial biomass and N uptake of maize canopies under different drought stress levels. However, the indices have also the potential to be used as indicators of plant water status and deficit stress as well as signaling the onset of plant dehydration (Claudio et al., 2006; Seelig et al., 2009). In particular, we observed that the index values decreased with increasing drought stress created through withholding irrigation during a certain period. These results agree with those of Yu et al. (2000), who reported that drought stress impacts significantly the physiological properties (e.g., water status) of maize, leading to a decrease in the values of the relevant physiological indices together with an increase in soil water stress.

Vegetation indices were developed to relate the reflectance of plant canopies with the characteristics necessary to estimate plant parameters such as dry matter production (Diker and Bausch, 2003; Hatfield et al., 2008). Many indices for green plants rely on wavelengths in the red edge (670 - 750 nm) and NIR (> 750 nm) region, which provide for good predictions of aboveground N-uptake of crops (Reusch et al., 2002; Reusch, 2003). These results agree with our finding that indices using wavelengths in this region showed the best relationships with aerial biomass and N uptake.

Differentiation of maize hybrids

The tropical maize hybrids used in this study showed differences with respect to the amount of aerial biomass and N uptake of plants. Through almost all sampling dates and irrigation treatments, hybrids were similarly classified in their amount of aerial biomass and N uptake. Significant differences in aerial biomass and N uptake were found only for hybrids with the maximum and minimum values. These results are similar to those reported by Edmeades et al. (1999), who found that differences in the biomass production of tropical maize experiencing drought versus those in well-watered environments were significant only for cultivars with the greatest and the least total aboveground biomass.

All hybrids are commercial hybrids planted in this region of Thailand. The hybrids were selected from major public and private seed agencies. Most hybrids have passed a wide area testing for several years before they were released to commercial uses. That is why the differences among hybrids were quite small. All hybrids are high and stable in yield under normal conditions and were not selected based on a range of drought tolerance for testing in this research. The aim was to detect differences among hybrids in terms of drought tolerance in order to help breeders in selection. The key point was that even if we detect small differences among the genotypes tested and the differences are consistent, a ranking of the hybrids would be possible and this method could help in the breeding process.

The longer the plants investigated here experienced drought stress and the less water that was available, the lower the aerial biomass and N uptake was, the availability of water obviously being essential for plant growth and development. Similar results have been reported in literature. Maize plants grown subjected to drought stress conditions have lower leaf-area indices, plant height, and biomass accumulation than under irrigated conditions (Edmeades et al., 1999; Soler et al., 2007), albeit with drought stress having a nonlinear impact on growth and development (Clay et al., 2006). Soler et al. (2007) also reported that different maize hybrids respond differently to soil water limitations, showing that genotypic differences in the response to drought stress exist and that management strategies should consider using cultivars that are adapted to particular environmental conditions (e.g., early maturity hybrids) in addition to optimal irrigation strategies. Indeed, the differentiation of maize cultivars is important in plant breeding for screening tests of drought tolerant plants (Inoue et al., 1993). Moreover, the fact that spectral reflectance is influenced by differences between crop varieties (Lammel et al., 2001), screening methods will need to account for this difference. In so doing, it is worth noting that the relationship of aerial biomass and N uptake with the spectral index R_{780}/R_{740} for each of the seven hybrids showed very high coefficients of determination (up to 0.88).

Although our results showed a strong relationship of index values with either of aerial biomass or N uptake, no direct differentiation of all varieties was possible. This could be caused by the high variability in both aerial biomass and N uptake between replicates making it difficult to detect differences between the hybrids. This variation might be due in part to the local agricultural standard (e.g., hand sowing and fertilizing) and/or the variability of the soil. It is possible that technological improvements (e.g., through a mechanization of the cultivation methods) could lead to better results with the potential to differentiate the hybrids even more precisely with even more controlled conditions. The high coefficients of variation of 18 % found on average for aerial biomass and 20 % for N uptake underline this suggestion. Previous results from other researchers have shown that under optimum cultivation conditions coefficient of variations may be close to 10% (Schmidt, 1995). However, the results presented in Table 5 support the assumption that comparable rankings of cultivars could be achieved based on destructive and most importantly sensor measurements. The advantage and great potential of our spectral based method is seen in its ability to deliver high throughput measurements of plant parameters because the sensor was mounted on a forklift and not handheld, which allows for a faster estimation of data on a large area. Furthermore, exact positioning of sensors over maize stands and constant angle of sensor view are considered to be advantageous. The success of highly mechanized sensing of plant phenotypes indicates the potential of combining automated sensing with tractor based mechanization to realize precision phenotyping techniques in plant breeding. Indeed, it has been shown that the spectral information obtained by a tractor-mounted crop scanning device enables the detection of the biomass and N status of plants on-the-go, an important consideration for establishing precision farming techniques (Schmidhalter et al., 2001) and likely for precision phenotyping.

CONCLUSIONS

We demonstrated that spectral indices can accurately estimate the aerial biomass and N uptake of seven different tropical maize hybrids at different drought stress levels, with the hybrids being mostly similarly classified in their amount of aerial biomass, N uptake and index values under both control and stress environments during the growing period. Overall, our results support that it may be possible to detect the aerial biomass and N uptake of maize hybrids at different irrigation treatments by high throughput phenotyping. In agreement with a previous suggestion (Schmidhalter, 2005), the assessment of biomass, N and water status of plants by high throughput sensing measurements may be seen as a promising technique for breeding purposes. Further testing with a larger genotype set is recommended. High throughput precision phenotyping combined with high throughput genotyping would provide researchers with the ability to gain detailed information of plant genotype characteristics that influence the plant phenotype, as well as the possibility of direct selection of the phenotype. This combination of methods could prove to be the major breakthrough in our attempts to explain the underlying genetic basis of complex plant traits such as drought tolerance (Campos et al., 2004; Montes et al., 2007; Richards et al., 2010).

Acknowledgement

The authors acknowledge the support of the German Federal Agency for Agriculture and Food (Project Nr. 2815303407).

References

Aparicio, N., D. Villegas, J. Al Araus, J. Casadesus, and C. Royo. 2002. Relationship between growth traits and spectral vegetation indices in durum wheat. Crop Sci. 42:1547-1555.

Camp, KH. 1996. Transpiration efficiency of tropical maize (*Zea Mays* L.). Ph.D. diss. Swiss Federal Institute of Technology Zürich, Switzerland.

Campos, H., M. Cooper, J. Habben, G. Edmeades, and J. Schussler. 2004. Improving drought tolerance in maize: A view from industry. Field Crop Res. 90:19-34.

Chappelle, E., J. McMurtrey III, F. Wood, and W. Newcomb. 1984. Laser-induced fluorescence of green plants. 2. LIF caused by nutrient deficiencies in corn. Appl. Optics. 23: 139-142.

Chaves, M., and B. Davies. 2010. Drought effects and water use efficiency: Improving crop production in dry environments. Funct. Plant Biol. 37:iii-vi.

Claudio, H., Y. Cheng, D. Fuentes, J. Gamon, H. Luo, W. Oechel, H. Qiu, A. Rahman, and D. Sims. 2006. Monitoring drought effects on vegetation water content and fluxes in chaparral with the 970 nm water band index. Remote Sens. Environ. 103:304-311.

Clay, D., K. Kim, J. Chang, S. Clay, and K. Dalsted. 2006. Characterizing water and nitrogen stress in corn using remote sensing. Agron. J. 98:579-587.

Diker, K., and C. Bausch. 2003. Potential use of nitrogen reflectance index to estimate plant parameters and yield of maize. Biosystems Eng. 85:437-447.

Edmeades, G., J. Bolanos, S. Chapman, H. Lafitte, and M. Bänziger. 1999. Selection improves drought tolerance in tropical maize populations: I. Gains in biomass, grain yield, and harvest index. Crop Sci. 39:1306-1315.

FAO. 2010. Food and Agriculture Organization of the United Nations. FAOSTAT. Available at http://faostat.fao.org/site/339/default.aspx. (verified 17 Feb. 2010).

Gates, D., H. Keegan, J. Schleter, and V. Weidner. 1965. Spectral properties of plants. Appl. Optics. 4:11-20.

Gitelson, A., Y. Gritz, and M. Merzlyak. 2003. Relationships between leaf chlorophyll content and spectral reflectance and algorithms for non-destructive chlorophyll assessment in higher plant leaves. J. Plant Physiol. 160:271-282.

Graeff, S., and W. Claupein. 2007. Identification and discrimination of water stress in wheat leaves (*Triticum aestivum* L.) by means of reflectance measurements. Irrigation Sci. 26:61-70.

Grant, L. 1987. Diffuse and specula characteristics of leaf reflectance. Remote Sens. Environ. 22:309–322.

Gutierrez, M., M.P. Reynolds, W. R. Raun, M.L. Stone, and A.R. Klatt. 2010. Spectral water indices for assessing yield in elite bread wheat genotypes under well-irrigated, water-stressed, and high temperature conditions. Crop Sci. 50, 197-214.

Hatfield, J., A. Gitelson, J. Schepers, and C. Walthall. 2008. Application of spectral remote sensing for agronomic decisions. Agron. J. 100:117–131.

Hu, Y., and U. Schmidhalter. 2005. Drought and salinity: A comparison of their effects on mineral nutrition of plants. J. Plant Nutr. Soil Sci. 168:541–549.

Inoue, Y., S. Morinaga, and M. Shibayama. 1993. Non-destructive estimation of water status of intact crop leaves based on spectral reflectance measurements. Jpn. J. Crop Sci. 62:462–469.

Lammel, J., J. Wollring, and S. Reusch. 2001.Tractor based remote sensing for variable nitrogen fertilizer application. In W.J. Horst et al. (ed.) Plant nutrition – Food security and sustainability of agro-ecosystems. 694–695. Kluver Academic Publishers. Netherlands.

Liu, L., C. Zhao, W. Huang, and J. Wang. 2003. Estimating winter wheat plant water content using red edge width. Int. J. Remote Sens. 25:3331–3342.

Mistele, B., and U. Schmidhalter. 2008. Spectral measurements of the total aerial N and biomass dry weight in maize using a quadrilateral-view optic. Field Crop Res. 106:94–103.

Mistele, B., and U. Schmidhalter. 2010a. A comparison of spectral reflectance and laser-induced chlorophyll fluorescence measurements to detect differences in aerial dry weight and nitrogen uptake of wheat. In R. Khosla (ed.) 10th International Conference in Precision Agriculture. Denver, Colorado July 18-21. 2010. CD-Rom. 14p.

Mistele, B., and U. Schmidhalter. 2010b. Tractor-based quadrilateral spectral reflectance measurements to detect biomass and total aerial nitrogen in winter wheat. Agron. J. 102:499–506.

Montes, J., A. Melchinger, and J. Reif. 2007. Novel throughput phenotyping platforms in plant genetic studies. Trends Plant Sci. 12:433–436.

Neidhart, B. 1994. Morphological and physiological responses of tropical maize (Zea mays L.) to pre-anthesis drought. Ph.D. diss. Swiss Federal Institute of Technology Zürich, Switzerland.

Passioura, J. 2007. The drought environment: Physical, biological and agricultural perspectives. J. Expt. Bot. 113–117.

Penuelas, J., J. Gamon, A. Fredeen, J. Merino, and C. Field. 1994. Reflectance indices associated with physiological changes in nitrogen- and water-limited sunflower leaves. Remote Sens. Environ. 48:135–146.

Penuelas, J., J. Pinol, R. Ogaya, and I. Filella. 1997. Estimation of plant water concentration by the reflectance Water Index WI (R900/R970). Int. J. Remote Sens. 18:2869–875.

Poss, J., W. Russell, and C. Grieve. 2006. Estimating yields of salt- and water-stressed forages with remote sensing in the visible and near infrared. J. Environ. Qual. 35:1060–1071.

Reusch, S., A. Link, and J. Lammel. 2002. Tractor-mounted multispectral scanner for remote field investigation. *In* P.C. Roberts (ed.) Proceedings of the 6th International Conference on Precision Agriculture, Minneapolis. ASA-CSSA-SSSA, Madison, WI, USA, pp. 1385–1393.

Reusch, S. 2003. Optimisation of oblique-view remote measurement of crop N-uptake under changing irradiance conditions. Precision Agriculture. Hydro Agri, Research Centre Hanninghof, Dülmen, Germany.

Richards, R., G. Rebetzke, M. Watt, A. Condon, W. Spielmeyer, and R. Dolferus. 2010. Breeding for improved water productivity in temperate cereals: Phenotyping, quantitative trait loci, markers and the selection environment. Funct. Plant Biol. 37:85–97.

Rodriguez, D., G. Fitzgerald, R. Belford, and L. Christensen. 2006. Detection of nitrogen deficiency in wheat from spectral reflectance indices and basic crop eco-biophysiological concepts. Aust. J. Agric. Res. 57:781–789.

Schlemmer, M., D. Francis, J. Shanahan, and J. Schepers. 2005. Remotely measuring chlorophyll content in corn leaves with differing nitrogen levels and relative water content. Agron. J. 97:106–112.

Schmid, W. 1995. Competition effects in silage maize surveys. (In German). Ph.D. diss. Technische Universität München, Germany.

Schmidhalter, U., J. Glas, R. Heigl, R. Manhart, S. Wiesent, R. Gutser, and E. Neudecker. 2001. Application and testing of a crop scanning instrument – field experiments with reduced crop width, tall maize plants and monitoring of cereal yield. *In* G. Grenier et al. (ed.) Proceedings of the 3rd European Conference on Precision Agriculture. Montpellier, France. pp. 953-958.

Schmidhalter, U. 2005. Sensing soil and plant properties by non-destructive measurements. Proceedings of the International Conference on Maize Adaption to Marginal Environments.

25th Anniversary of the Cooperation between Kasetsart University and Swiss Federal Institute of Technology. March 6-9. Nakhon Ratchasima, Thailand.

Seelig, H., W. Adams III, A. Hoehn, L. Stodieck, D. Klaus, and W. Emery. 2008a. Extraneous variables and their influence on reflectance-based measurements of leaf water content. Irrigation Sci. 26:407–414.

Seelig, H., A. Hoehn, L. Stodieck, D. Klaus, W. Adams III, and W. Emery. 2008b. Relations of remote sensing leaf water indices to leaf water thickness in cowpea, bean, and sugarbeet plants. Remote Sens. Environ. 112:445–455.

Seelig, H., A. Hoehn, L. Stodieck, D. Klaus, W. Adams III, and W. Emery. 2009. Plant water parameters and the remote sensing R1300/R1450 leaf water index: controlled condition dynamics during the development of water deficit stress. Irrigation Sci. 27:357–365.

Soler, C., G. Hoogenboom, P. Sentelhas, and A. Duarte. 2007. Impact of water stress on maize grown off-season in a subtropical environment. J. Agron. Crop Sci. 193:247–261.

Thoren, D., and U. Schmidhalter. 2009. Nitrogen status and biomass determination of oilseed rape by laser-induced chlorophyll fluorescence. Eur. J. Agron. 30:238–242.

Tilling, A., G. O'Leary, J. Ferwerda, S. Jones, G. Fitzgerald, D. Rodriguez, and R. Belford. 2007. Remote sensing of nitrogen and water stress in wheat. Field Crop Res. 104:77–85.

Yu, G., T. Miwa, K. Nakayama, N. Matsuoka, and H. Kon. 2000. A proposal for universal formulas for estimating leaf water status of herbaceous and woody plants based on spectral reflectance properties. Plant Soil. 227:47–58.

Publication II

High throughput phenotyping of canopy water mass and canopy temperature in well-watered and drought stressed tropical maize hybrids in the vegetative stage

ABSTRACT

The high throughput determination of the water status of maize (*Zea mays* L.) in precision agriculture presents numerous benefits, but also shows the potential for improvement. On the former count, the differentiation of maize hybrids could be used in screening drought tolerance in plant breeding, whereas, on the latter count, the monitoring of plant water status by non-destructive high-throughput sensing carried out on GPS based vehicles could enable the fast evaluation of various traits over a large area, improving the management decisions of farmers. The aim of this study was to assess the ability to measure the canopy water mass (CWM; amount of water in kg m^{-2}) of several tropical maize hybrids using high throughput sensing. Experimental field trials were conducted in Thailand (National Corn and Sorghum Research Center) in the years 2007-2009, where seven hand sown tropical high yield hybrids were analyzed under four furrow irrigation treatments. High throughput canopy reflectance and thermal radiance measurements, as well as biomass samplings were done on a regular basis until flowering. Both a large number of spectral indices from literature and newly developed for this study were validated. Selected spectral indices and IR-temperature were highly correlated with CWM and able to show the different drought stress levels. Several indices showed global coefficients of determination of over 0.70 and it was possible to differentiate and classify the hybrids into three consistent groups (above, below, or average performance) under control and stress environments. The results of this study show that it is indeed possible to both detect CWM and discriminate between groups of hybrids using non-destructive high throughput phenotyping, and that this technology presents a potentially useful application for breeding in the future.

Keywords: Breeding; Drought tolerance; Optical sensor; Phenotyping; Precision Agriculture; Reflectance; Screening; Spectral Index; Thermometry.

1. Introduction

The combination of climate change and shortage of water suitable for agriculture raises the potential for drought events (Chaves and Davies, 2010; Passioura, 2007). Thus, two key goals in agricultural production should be the adjustment of crop phenology to drought environments as well as of agronomic practices to a more effective use of water. Both developments would help counteract the threat of increased occurrence of droughts in the future by leading to an improved employment of limiting water supplies in agriculture and an enhanced drought resistance of crops (Blum, 2009).

The measurement of the water status of maize provides crucial information to understand its interaction with the environment. For instance, it is known that a lack of water and nitrogen negatively affects the quantity of chlorophyll produced in plants as well as cell turgidity. With increasing water deficit, plants appear increasingly wilted and the photochemical activity of chlorophyll is lessened.

Farmers could profit greatly by basing management decisions on remote sensing techniques (Osborne et al., 2002) designed to detect variations in plant development within a field (Diker and Bausch, 2003) through fast and accurate measurements of factors that impede crop growth and limit biomass production (Poss et al., 2006). For instance, it is known that water and N stresses have a negative impact on crop yield because N uptake is dependent on soil and plant water status (Hu and Schmidhalter, 2005). Spectral measurements would also be very useful in plant breeding contexts for screening drought tolerance of crops (Inoue et al., 1993; Schmidhalter, 2005).

The foundation of remote sensing methods lies with the fact that the water status of plants directly influences the intercellular air spaces and cell turgidity, such that the leaf cell structure affects how light will be absorbed, transmitted, or reflected by leaves (Schlemmer et al., 2005). Reflection occurs mainly in the layer of spongy mesophyll cells as well as in the layer of palisade cells due to differences between cell walls and intercellular air spaces (Seelig et al., 2008b). Drought stress, by causing alterations in leaf cell structure and composition (e.g., changing the properties of connections between cell walls and air spaces, cell sizes and shapes, and/or cell wall composition and structure), will therefore result in variation in spectral reflectance values (Grant, 1987; Liu et al., 2003; Penuelas et al., 1994). In particular, changes in the leaf internal structure when the amount of water in leaves is reduced influence spectral reflectance in the red edge (680 – 740 nm) and near infrared (740 – 940 nm) regions (Liu et al., 2003; Tilling et al., 2007) as well as increase reflectance in the 400 - 1300 nm region (Graeff and Claupein, 2007; Inoue et al., 1993).

The near-infrared region (NIR) leaf reflectance spectrum presents several water absorption bands that offer in combination the possibility to measure the leaf water content. For example, leaf water indices can be obtained by dividing reflectance at a spectral region that is at best only weakly absorbed by water by that at a region that is strongly absorbed by water, thereby measuring the absolute amount of absorbing water that is present within the path of light reflected from leaves directly (Seelig et al., 2009). Spectral indices derived from real-time multispectral reflectance information have the potential to be used as indicators of plant water status and deficit stress and to signal the onset of plant dehydration when the water deficit stress becomes too severe for a plant to cope with (Claudio et al., 2006; Seelig et al., 2008a), leading to a non lethal drought stress allowing recovery. They can also be used to determine canopy water content (CWC; Colombo et al., 2008; Rollin and Milton, 1998) and to quickly determine parameters including plant water content (PWC; Penuelas et al., 1997), relative water content (RWC; Yu et al., 2000) and equivalent water thickness (EWT; Seelig et al., 2008a). In addition, the plant water status can also be estimated through the canopy temperature using infrared thermometry (Hunt and Rock, 1989; Peters and Evett, 2007).

The combination of high throughput phenotyping joined with high throughput genotyping could lead to a major breakthrough in our understanding of the fundamental genetic basis of complex plant traits like drought tolerance (Edmeades et al., 2004). Unfortunately, our ability to characterize gene-to-phenotype relationships for drought tolerance in maize is hampered by the comparative delay in developing methods to efficiently measure plant phenotypes for important traits, an area that has lagged behind the advances that have been made in plant genotyping. With information obtained from high throughput precision phenotyping, the researcher could get detailed information of plant phenotypes determined by various genotypes from a plant breeding population and have the possibility to understand this relation (Campos et al., 2004; Thoren and Schmidhalter, 2009).

Accordingly, the most important restriction in improving selection methods for drought stress environments is the deficit in fast and precise measurements of the phenotype. Indeed, we argue that enhancing the speed and precision of phenotyping plant traits will prove to be more advantageous for effecting progress in plant breeding than will additional advances in molecular technologies. The evaluation of large populations in plant breeding requires that early selection and effective reduction be conducted as quickly and as cheaply as possible (Richards et al., 2010). To potentially act on these objectives, it is important that remote sensing researchers completely understand the potential of their methodology, which, in turn, requires an amelioration of the basic knowledge of the agronomic information content of remote sensing data and algorithmic innovations for analysis (Hatfield et al., 2008).

To further this goal, the aim of our study was to determine the canopy water mass (CWM) of several tropical maize hybrids experiencing different drought stress treatments using high throughput sensing measurements, with the possibility of developing a technique that will prove to be potentially useful for breeding purposes in the future. Therefore, spectral indices used in the literature to detect parameters including PWC, RWC, EWT as well as plant dry matter, nitrogen and chlorophyll were used in combination with newly developed indices to assess our ability of accurate measure CWM of the maize hybrids using remote sensing techniques.

2. Material and Methods

2.1 Experimental design

The experiment was conducted at the National Corn and Sorghum Research Center of Thailand (101.3 E long; 14.6 N lat; 380 m alt), located 155 km northeast of Bangkok. The climate in this tropical region is influenced by a rainy and dry season, the latter occurring from November to February and being characterized by no rainfall. Trials were carried out in the dry seasons in the years 2007/08 and 2008/09. The average temperature (November 1 to January 31) in the first year was 24 °C and 22.5 °C in the second year. The soil of the research station is classified as reddish brown lateritic soil with moderately high water permeability (Neidhart, 1994).

Seven maize hybrids, four irrigation treatments and four replications were arranged as a randomized block design with a total of 112 plots. Each plot was 10 m in length and consisted of 10 rows separated by 0.75 m, with a distance of 0.20 m between the plants in each row. The seven tropical high yield maize hybrids were selected from different sources (see Table 1) and were hand sown. The different drought stress levels were obtained by four irrigation treatments: a control treatment with weekly irrigation and three stress treatments withholding irrigation for different periods of time (Table 2). Sprinkler irrigation (40 mm per application) was applied until 3 weeks after sowing, followed thereafter by furrow irrigation (65 mm per application). Irrigation methods on this location were applied based on long-term experience of the management staff and backed up by tensiometric information. Fertilization was applied as basal fertilizer with two components (N+P) at a rate of 20 kg N ha^{-1} and urea as side-dress at a rate of 115 kg N ha^{-1}, leading to a total fertilization rate of 135 kg N ha^{-1}, 25 kg P ha^{-1} and 0 kg K ha^{-1}. Cultivation methods (e.g. thinning and hand weeding) were done following local technical recommendations and there was no need to do any pest management.

Table 1
Tropical maize hybrids used in this study.

No.	Hybrid	Source
1	Pac 224	Pacific Seeds (Thai) Limited
2	NK 40	Syngenta Seeds Limited
3	30Y87	Pioneer Hi-Bred (Thailand) Co., Ltd.
4	CP AAA Super	Bangkok Seeds Industry Co., Ltd.
5	DK 979	Monsanto Seeds (Thailand) Limited
6	NS 2	Nakhon Sawan Field Crops Research Center
7	Suwan 4452	National Corn and Sorghum Research Center

Table 2
Irrigation treatments, important dates and biomass samplings with days after sowing and vegetation stages of maize for the experiments in 2007/08 and 2008/09.

Irrigation treatment	Experiment 2007/08	Experiment 2008/09
Control (C)	weekly irrigation	weekly irrigation
early vegetative stress (ES)	no irrigation: 07 Dec. 2007 - 03 Jan. 2008	no irrigation: 13 Dec. 2008 - 08 Jan. 2009
late vegetative stress (LS)	no irrigation: 22 Dec. 2007 - 17 Jan. 2008	no irrigation: 20 Dec. 2008 - 15 Jan. 2009
pre-flowering stress (PS)	no irrigation: 29 Dec. 2007 - 24 Jan. 2008	no irrigation: 27 Dec. 2008 - 29 Jan. 2009
Important dates		
sowing date	13 Nov. 2007	18 Nov. 2008
begin of flowering	18 Jan. 2008	29 Jan. 2009
1. biomass sampling	35 DAS (V6)	35 DAS (V6)
2. biomass sampling	42 DAS (V7)	49 DAS (V8)
3. biomass sampling	56 DAS (V9)	56 DAS (V9)
4. biomass sampling	63 DAS (V11)	63 DAS (V11)

2.2 Plant and soil water status measurements

Biomass samplings were done regularly until flowering to determine the amount of water in kg ha^{-1} contained in the above ground biomass (= CWM) of the maize plants. At each harvest, 30 plants were cut from 6 m in the inner rows (i.e., no border plants were selected) and their fresh weights were determined. The plant material was chopped and a representative subsample was weighed, oven dried at 100 °C for 3 days and then reweighed. Grain yield was determined on two rows in the middle of April.

Tensiometers were installed at depths of 20, 40, 60, 80 and 100 cm to monitor the soil matric potential and to control the drying out of the soil during the drought stress periods. All measurements were done regularly until flowering and were replicated six times in the control, early vegetative stress and late vegetative stress treatments.

2.3 Spectral and thermal radiance measurements

Canopy reflectance measurements were done until flowering, as simultaneously to the biomass samplings as possible and always at the same time (between 11:00 and 13:00). The sensor system was mounted on a forklift, with the frame being placed on the left side of the carrier, measuring exactly over the maize canopy and with the possibility to vary the height of the sensor enabling measurements of fully developed maize canopies with plant heights of 2.5 m. Driving lanes next to the plots allowed a fast and non-destructive possibility to measure the maize canopies through the entire growing period. The sensor with modified electronics (tec5, Oberursel, Germany) contained two units of a Zeiss MMS1 silicon diode array spectrometer that measure reflectance and incident radiation at the same time. The optics were mounted with an angle of 50° on the corners of the frame, leading to an oblique and oligo view optic. One component was connected to a four-in-one light fiber to create an optical mixed signal of the canopy reflection of one area from four different directions, with the spectrometer analyzing the reflected radiation in 256 spectral channels with a detection range from 300 to 1000 nm and a bandwidth of 3.3 nm. The second component was connected to a diffuser to measure the global radiation to compensate for different light conditions, with the associated spectrometer having a detection range from 1000 to 1700 nm and a bandwidth of 6 nm (Mistele and Schmidhalter, 2008).

Canopy temperature was measured with an infrared (IR) thermometer (KT15D, Heitronics Infrarot Messtechnik GmbH, Wiesbaden, Germany) that was also mounted on the sensor system so that the measurements were done simultaneously with those of canopy reflectance. All spectral and thermal radiance measurements were co-registered with corresponding GPS data.

2.4 Spectral indices

Spectral indices, often calculated with the reflectance values from two wavelengths, were used to detect the relationship of agronomical traits of plant canopies with their reflectance. Several indices have been described that relate successfully to plant biomass, nitrogen and chlorophyll (Mistele and Schmidhalter, 2008; Rodriguez et al., 2006; Tilling et al., 2007) as well as to plant water content (Claudio et al., 2006; Penuelas et al., 1997) and to relative water content (Inoue et al., 1993; Yu et al., 2000). To evaluate the sensor system as well as to assess its ability to measure CWM, a wide range of these spectral indices from the literature as well as newly developed indices were tested. A detailed overview of the spectral indices is given in Table 3.

Table 3

Spectral indices evaluated in this study presented with name, formula, parameter with primary sensitivity, corresponding plant species, analyzed plant level, location, and citation. The detected parameters were CWM = canopy water mass (kg/m^2); PWC = plant water content (%); RWC = relative water content (%); WC = water content (mg/cm^2); LWC = leaf water content (g/m^2); EWT = equivalent water thickness (µm); DM = dry matter yield (kg/ha); N = total aerial nitrogen (kg/ha); %N = nitrogen content (%); Chl = chlorophyll content (mol/m^2); %Chl = Chlorophyll concentration (%); LAI = leaf area index; YLWS & YLNS = Yield losses (kg/ha) due to water and N stress, respectively.

Index	Formula	Parameter	Species	Plant level	Location	Author
CWMI I (850/725)	R_{850}/R_{725}	CWM	maize	canopy	field	this work
CWMI II (890/715)	R_{890}/R_{715}	CWM	maize	canopy	field	this work
CWMI III (980/715)	R_{980}/R_{715}	CWM	maize	canopy	field	this work
760/730	R_{760}/R_{730}	N, DM	maize	canopy	field	Mistele and Schmidhalter (2010)
SR (simple ratio)	R_{900}/R_{680}	LAI	durum wheat	canopy	field	Aparicio et al. (2002); Poss et al. (2006)
WBI/NDVI	$[R_{900}/R_{970}]/[(R_{800}-R_{680})/(R_{800}+R_{680})]$	PWC	trees, shrubs, grass	canopy	field	Claudio et al. (2006); Penuelas et al. (1997)
WBI (Water Band Index)	R_{900}/R_{970}	PWC	trees, shrubs, grass	canopy	field	Claudio et al. (2006); Penuelas et al. (1997)
NDWI (Normalized Difference Water Index)	$(R_{840}-R_{1650})/(R_{840}+R_{1650})$	YLWS & YLNS	maize	canopy	field	Clay et al. (2006)
970/1450	R_{970}/R_{1450}	water bands	soybean	leaf	pot	Dallon 2003 (wavelength)
1100/1200	R_{1100}/R_{1200}	RWC, WC	maize, peanut, soybean, wheat	leaf	field	Inoue et al. (1993)
1100-1200	$R_{1100}-R_{1200}$	RWC, WC	maize, peanut, soybean, wheat	leaf	field	Inoue et al. (1993)
800/1650	R_{800}/R_{1650}	RWC, WC	maize, peanut, soybean, wheat	leaf	field	Inoue et al. (1993)
1200/1430	R_{1200}/R_{1430}	RWC, WC	maize, peanut, soybean, wheat	leaf	field	Inoue et al. (1993)
800-1200	$R_{800}-R_{1200}$	RWC, WC	maize, peanut, soybean, wheat	leaf	field	Inoue et al. (1993)
1650/1430	R_{1650}/R_{1430}	RWC, WC	maize, peanut, soybean, wheat	leaf	field	Inoue et al. (1993)
NIR/NIR	R_{780}/R_{740}	N, DM	maize	canopy	field	Mistele and Schmidhalter (2008)
NIR/red	R_{780}/R_{700}	N, DM	maize	canopy	field	Mistele and Schmidhalter (2008)
NIR/green	R_{780}/R_{550}	N, DM	maize	canopy	field	Mistele and Schmidhalter (2008)
NDVI (Normalized Difference Vegetation Index)	$(R_{800}-R_{680})/(R_{800}+R_{680})$	N, DM, %N, %Chl	maize, wheat	canopy	field	Mistele and Schmidhalter (2008); Tilling et al. (2007)
440/685	R_{440}/R_{685}	RWC, Chl	maize	leaf	pot	Schlemmer et al. (2005)
525/685	R_{525}/R_{685}	RWC, Chl	maize	leaf	pot	Schlemmer et al. (2005)
YCAR	R_{600}/R_{680}	RWC, Chl	maize	leaf	pot	Schlemmer et al. (2005)
OCAR	R_{630}/R_{680}	RWC, Chl	maize	leaf	pot	Schlemmer et al. (2005)
1300/1450	R_{1300}/R_{1450}	EWT	tree, bean	leaf	field, pot	Seelig et al. (2008a)
NDRE (Normalized Difference Red Edge)	$(R_{790}-R_{720})/(R_{790}+R_{720})$	%N, %Chl	wheat	canopy	field	Rodriguez et al. (2006); Tilling et al. (2007)
1483/1650	R_{1483}/R_{1650}	RWC, LWC	maize, soybean, tuliptree, vibumum	leaf	pot	Yu et al. (2000)
1121/1430	R_{1121}/R_{1430}	RWC, LWC	maize, soybean, tuliptree, vibumum	leaf	pot	Yu et al. (2000)
1100/1430	R_{1100}/R_{1430}	RWC, LWC	maize, soybean, tuliptree, vibumum	leaf	pot	Yu et al. (2000)
1430/1650	R_{1430}/R_{1650}	RWC, LWC	maize, soybean, tuliptree, vibumum	leaf	pot	Yu et al. (2000)

2.5 Statistical Analysis

Statistical analysis was conducted to determine the effect of the maize hybrid and different irrigation treatments, both individually and in combination, on the relationship of CWM with the different spectral indices. Curvilinear models (quadratic) in Microsoft Excel 2003 (Microsoft Inc., Seattle, WA, USA), as well as analysis of variance (ANOVA) and general linear model (GLM) analysis in SPSS 16 (SPSS Inc., Chicago, USA) were used to establish relationships. Finally, hybrids were classified as having above, below or average performance using the method of Worku et al. (2007). Here, hybrid performance under the control treatment was plotted against that of the early and late vegetative stress treatments, making it possible to distinguish between better or less well performing hybrids. Figures were selected to exemplify the results and show only a few sampling dates representing the total evaluation.

3. Results

3.1 Soil water status

Soil matric potentials in the experimental year 2007/08 for the different irrigation treatments are shown in figure 1; the tensiometer results in the experimental year 2008/09 were similar (data not presented). In the control treatment (Fig. 1a), the weekly irrigation resulted in virtually no drying of the uppermost 20 cm of the soil. By contrast, the early vegetative stress treatment displayed a strong drying out of the soil in the uppermost 20 cm and a slight drying out in the 21-40 cm layer, indicating the major rooting depth and water uptake (Fig 1b). The late vegetative stress treatment (Fig. 1c) showed a strong drying out of the soil down to a depth of 60 cm and a slight drying out in the layer 61 - 100 cm, revealing that rooting depth and activity probably descend even further down.

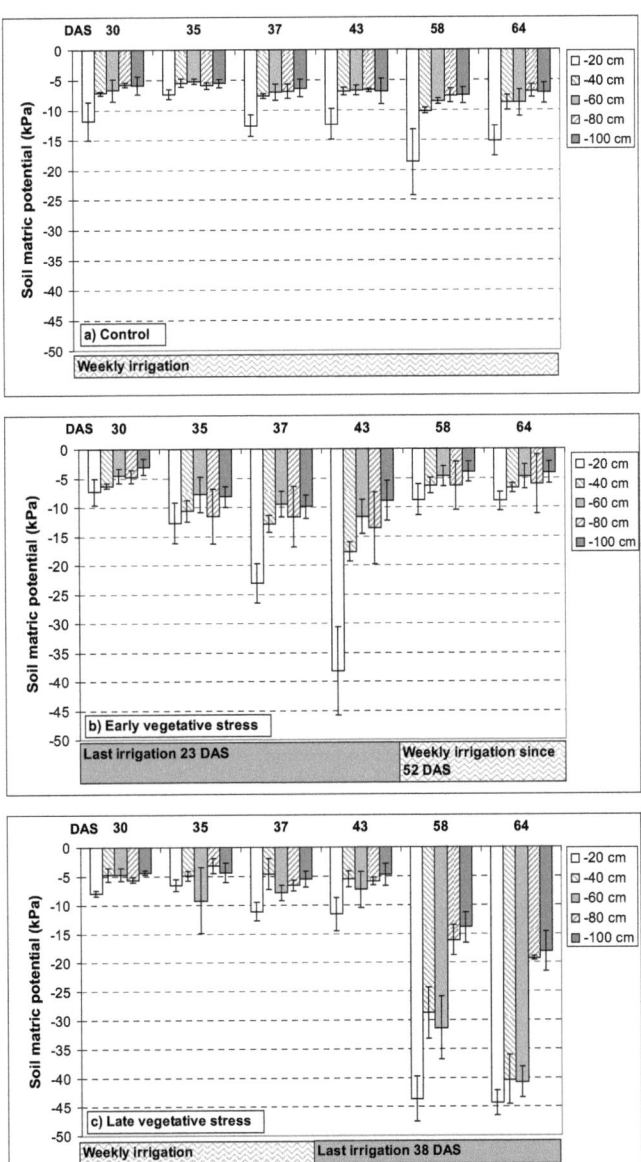

Fig. 1. Soil matric potential measured at different depths using tensiometers for the three irrigation treatments control (a), early vegetative stress (b) and late vegetative stress (c) in the experimental year 2007/08. No data are available for the pre-flowering stress. Error bars indicate standard deviations.

3.2 Influence of irrigation treatment and hybrid on canopy water mass

A full analysis of variance showed that all investigated effects were highly significant, with highest F-values being observed for sampling date, followed by year, irrigation treatments and cultivar effects. Interaction effects were also highly significant except for cultivar*irrigation treatment*sampling date, cultivar*irrigation treatment*year, cultivar*sampling date*year and cultivar*irrigation treatment*sampling date*year. No two-way interactions were found. In the subsequent analysis, years and sampling dates were evaluated separately (Table 4). Years were evaluated separately because of the cooler weather in the second experimental year, leading to smaller plants during the vegetation period. The purpose of the analysis was to show possible differences among hybrids at specific sampling dates, hence they were analyzed individually, like plant breeders also rating plants at specific dates during the vegetation period.

Irrigation treatments and tropical maize hybrids had individually a significant influence on the CWM for all measurements in the GLM analysis for the experimental years 2007/08 and 2008/09 (Table 4), except for the irrigation treatment 35 DAS (2008/09). There was no influence on the interaction of hybrids with irrigation treatments.

For all sampling dates and irrigation treatments, varieties were similarly classified in their amount of CWM. A statistically based grouping of the hybrids is presented in Table 5. Significant differences in CWM were found only for hybrids with the maximum and minimum values in the experimental years 2007/08 and 2008/09. However, for grain yield no consistent grouping was possible.

Table 4

Results from an analysis of variance measuring the influence of each of hybrid and irrigation treatment on canopy water mass (CWM) in the experimental years 2007/08 and 2008/09.

Experiment 2007/08		CWM		
Days after sowing	Parameter	df	F value	Sig.
35	Hybrid	6	13.6	***
	Irrigation treatment	1	4.9	**
	Hybrid*Irrigation treatment	12	0.2	
42	Hybrid	6	10.8	***
	Irrigation treatment	2	29.2	***
	Hybrid*Irrigation treatment	12	0.4	
56	Hybrid	6	17.2	***
	Irrigation treatment	3	84.4	***
	Hybrid*Irrigation treatment	18	0.6	
63	Hybrid	6	21.0	***
	Irrigation treatment	3	94.8	***
	Hybrid*Irrigation treatment	18	0.6	

Experiment 2008/09		CWM		
Days after sowing	Parameter	df	F value	Sig.
35	Hybrid	6	5.4	***
	Irrigation treatment	2	0.1	
	Hybrid*Irrigation treatment	12	0.5	
49	Hybrid	6	6.0	***
	Irrigation treatment	3	11.1	***
	Hybrid*Irrigation treatment	18	0.5	
56	Hybrid	6	3.5	**
	Irrigation treatment	3	15.2	***
	Hybrid*Irrigation treatment	18	0.3	
63	Hybrid	6	4.0	**
	Irrigation treatment	3	16.3	***
	Hybrid*Irrigation treatment	18	0.5	

*Significant at P≤0.05.
**Significant at P≤0.01.
***Significant at P≤0.001.

Table 5

Results from an analysis of variance of canopy water mass and index R_{780}/R_{740} of seven maize hybrids (average of all irrigation treatments) for the experimental years 2007/08 and 2008/09. Letters (a, b, c) indicate groups of hybrids that were significantly different from each other at the 0.05 level (SNK test). The hybrids were Pac 224 (1), NK 40 (2), 30Y87 (3), CP AAA Super (4), DK 979 (5), NS 2 (6) and Suwan 4452 (7).

Year 2007/08						Year 2008/09					
Hybrid No.	All harvests	DAS				Hybrid No.	All harvests	DAS			
	CWM	35	42	56	63		CWM	35	49	56	63
7	a	a	a	a	a	7	a	a	a	a	a
6	a	a	a	a, b	a	6	a	a, b	a, b	a	a
4	a, b	b	b	a, b, c	a, b	4	a	b, c	a, b	a, b	a
1	a, b	b	b	a, b, c	a, b	2	a, b	b, c	a, b	a, b	a, b
3	a, b	b	b	b, c	a, b	3	a, b	b, c	b, c	a, b	a, b
2	b	b	b	c	c	1	a, b	b, c	b, a, c	b	b
5	b	b	b	c	c	5	b	c	c	b	b
	R_{780}/R_{740}	36	43	58	64		R_{780}/R_{740}	37	50	58	64
6	a	a	a	a	a	6	a	a, b	a	a	a
7	b	a, b	a, b	a, b	a, b	4	a	b, c	a	a	a
3	b, c	b, c	b, c	b	a, b	2	a	a, b	a	a	a
1	b, c	b, c	b, c	b	a, b	7	a	a	a	a	a
4	b, c	b, c	b, c	b	a, b	1	a, b	c	a, b	a	a
2	b, c	c	c	b	b	3	a, b	c	a, b	a	a
5	c	c	c	b	b	5	b	c	b	a	a

3.3 Separation of well-watered and drought stressed treatments

An overview of the relationship between the CWM and all spectral indices tested is given in Table 6. The six best indices showed average coefficients of determination (R^2) exceeding 0.70 over all experiments from 2007 to 2009. A more detailed view over the entire growing period (Fig. 2) reveals the generally good relationship between CWM and the best indices, with the relationship being better for the first experimental year.

Table 6
Correlation of spectral indices with canopy water mass (kg/m^2). Coefficients of determination (R^2) represent mean values of 16 measurements from 2007 to 2009.

Index	Formula	R^2
CWMI I (850/725)	R_{850}/R_{725}	0.72
NIR/NIR	R_{780}/R_{740}	0.72
760/730	R_{760}/R_{730}	0.72
CWMI II (890/715)	R_{890}/R_{715}	0.71
NDRE	$(R_{790}-R_{720})/(R_{790}+R_{720})$	0.71
CWMI III (980/715)	R_{980}/R_{715}	0.71
NIR/red	R_{780}/R_{700}	0.68
1100/1200	R_{1100}/R_{1200}	0.68
1100-1200	$R_{1100}-R_{1200}$	0.67
SR	R_{900}/R_{680}	0.65
NIR/green	R_{780}/R_{550}	0.64
NDVI	$(R_{800}-R_{680})/(R_{800}+R_{680})$	0.63
440/685	R_{440}/R_{685}	0.63
WBI	R_{900}/R_{970}	0.61
1483/1650	R_{1483}/R_{1650}	0.61
525/685	R_{525}/R_{685}	0.58
WBI/NDVI	$[R_{900}/R_{970}]/[(R_{800}-R_{680})/(R_{800}+R_{680})]$	0.58
NDWI	$(R_{840}-R_{1650})/(R_{840}+R_{1650})$	0.58
800/1650	R_{800}/R_{1650}	0.57
1300/1450	R_{1300}/R_{1450}	0.56
YCAR	R_{600}/R_{680}	0.51
970/1450	R_{970}/R_{1450}	0.50
OCAR	R_{630}/R_{680}	0.46
1121/1430	R_{1121}/R_{1430}	0.45
1100/1430	R_{1100}/R_{1430}	0.45
1200/1430	R_{1200}/R_{1430}	0.42
800-1200	$R_{800}-R_{1200}$	0.39
1430/1650	R_{1430}/R_{1650}	0.37
1650/1430	R_{1650}/R_{1430}	0.33

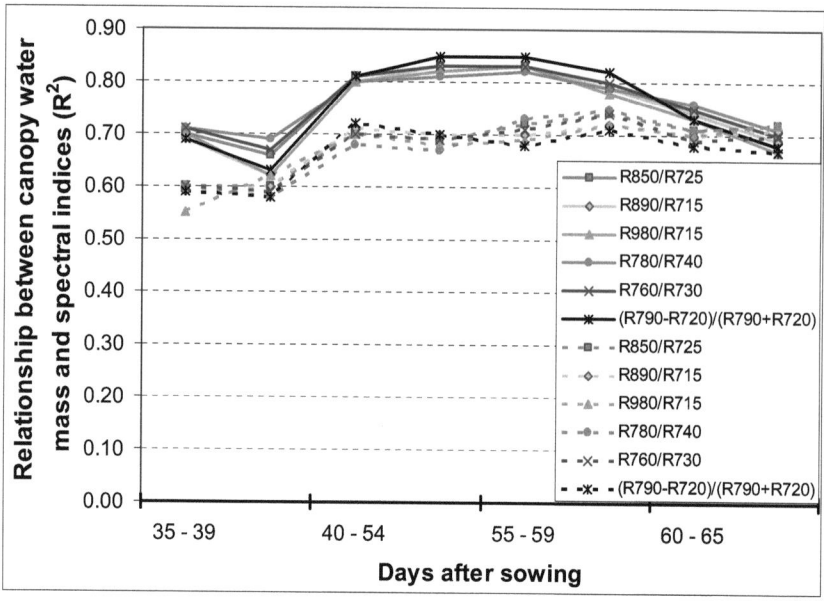

Fig. 2. Coefficient of determination (R^2) between canopy water mass and spectral indices for the experimental years 2007/08 (continuous lines) and 2008/09 (dotted lines). Measurement periods are indicated for the respective spectral recordings in the two years. The indices were CWMI I (R_{850}/R_{725}), NIR/NIR (R_{780}/R_{740}), 760/730 (R_{760}/R_{730}), CWMI II (R_{890}/R_{715}), NDRE (($R_{790}-R_{720})/(R_{790}+R_{720})$) and CWMI III ($R_{980}/R_{715}$).

Given that the analysis of variance indicated the significant influence of irrigation treatment on CWM (see above), we attempted to use spectral indices to detect differences between the irrigation treatments. As an example we present the relationships between CWM and the spectral index R_{850}/R_{725} 57 days after sowing in the experimental year 2007/08 (Fig. 3). A close relationship over all irrigation treatments was apparent (overall R^2 = 0.83), with lower R^2 values being found for the individual treatments, especially for the pre-flowering stress. Different stress treatments could be distinguished through the height of the index values: the more irrigation that was provided, the higher was the index value over all measurements.

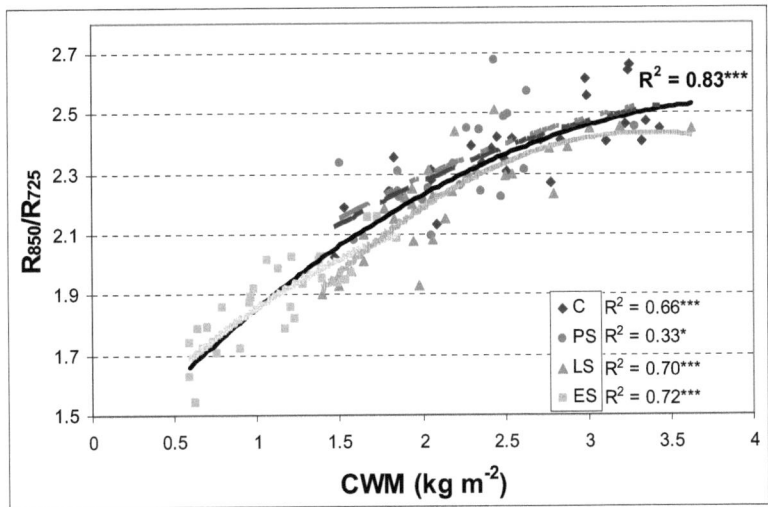

Fig. 3. Relationship of canopy water mass and spectral index R_{850}/R_{725} at 57 days after sowing in the experimental year 2007/08. Coefficients of determination averaged over all hybrids and irrigation treatments (R^2 = 0.83) and for individual irrigation treatments (C = control, PS = pre-flowering stress, LS = late vegetative stress and ES = early vegetative stress) are presented. *Significant at P≤0.05, **Significant at P≤0.01, ***Significant at P≤0.001.

The relationship between CWM and canopy IR-temperature is presented in Fig. 4, where the coefficients of determination over all irrigation treatments (R^2 = 0.65) and for each treatment individually were roughly similar. Again, the different stress levels could be differentiated by thermal radiance measurements of the canopy: the more water that was available for the plants, the lower was the canopy temperature over all measurements.

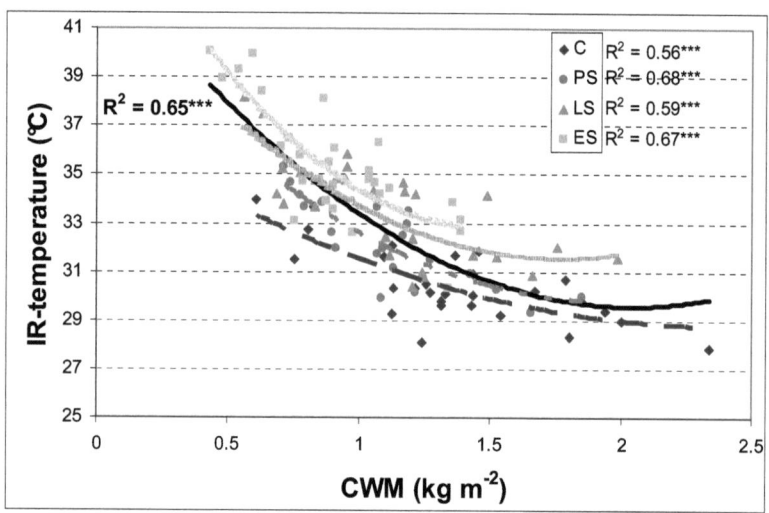

Fig. 4. Relationship of canopy water mass and IR-temperature at 50 days after sowing in the experimental year 2008/09. Coefficients of determination averaged over all hybrids and irrigation treatments (R^2 = 0.65) and for individual irrigation treatments (C = control, PS = pre-flowering stress, LS = late vegetative stress and ES = early vegetative stress) are presented. *Significant at P≤0.05, **Significant at P≤0.01, ***Significant at P≤0.001.

To detect the water status of the maize plants through non-destructive, high-throughput sensing measurements, IR-temperature was related to the spectral index R_{760}/R_{730}, as shown in Fig. 5 for the measurement 50 days after sowing in the experimental year 2008/09. Coefficients of determination ranged between 0.65 and 0.78, and clearly differentiated the different irrigation treatments. The longer the plants experienced drought stress from withholding water, the lower was the index value and the higher was the temperature.

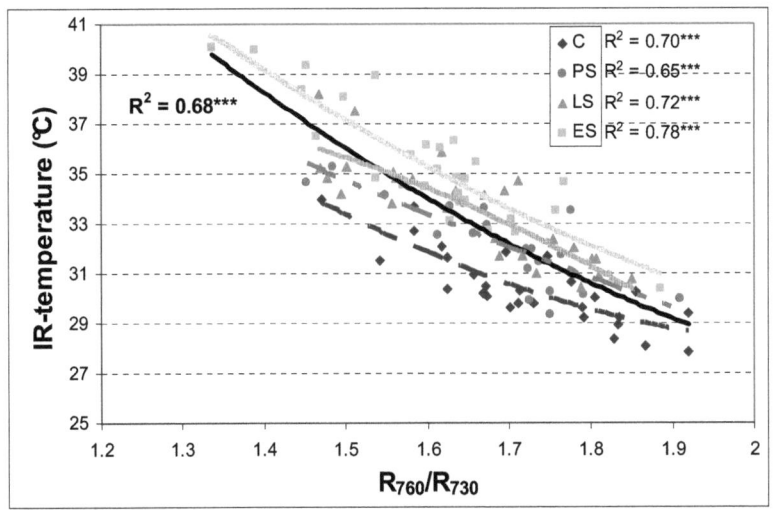

Fig. 5. Relationship of spectral index R_{760}/R_{730} and IR-temperature at 50 days after sowing in the experimental year 2008/09. Coefficients of determination averaged over all hybrids and irrigation treatments (R^2 = 0.68) and for individual irrigation treatments (C = control, PS = pre-flowering stress, LS = late vegetative stress and ES = early vegetative stress) are presented. *Significant at $P \leq 0.05$, **Significant at $P \leq 0.01$, ***Significant at $P \leq 0.001$.

3.4 Differentiation of CWM of the tropical maize hybrids by spectral indices

Using the method of Worku et al. (2007), the performance of hybrids was classified into three consistent groups (above, below or average performance) throughout the growing period (Fig. 6). Using empirically derived values of CWM, hybrids 6 and 7 were classified as having above average performance; hybrids 1, 3 and 4 as average performance; and hybrids 2 and 5 as below average performance of the hybrids (Fig. 6a and 6b). Given that the same results were obtained using the spectral index R_{780}/R_{740} (Fig. 6c), CWM and the index R_{780}/R_{740} of the control treatment were plotted against each other, showing the same classification for the hybrids (Fig. 6d). Analogous results were achieved for all measurements and relationships throughout the entire experiment from 2007 to 2009 (data not shown).

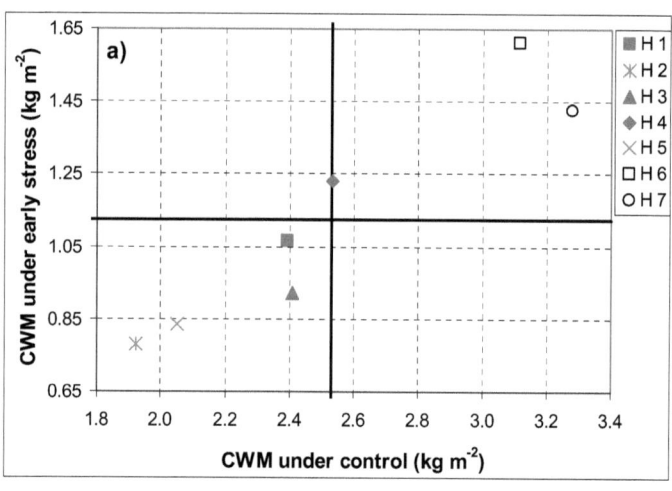
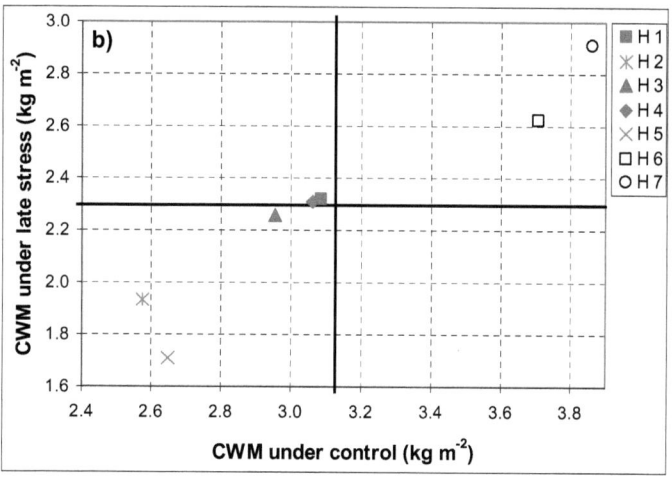

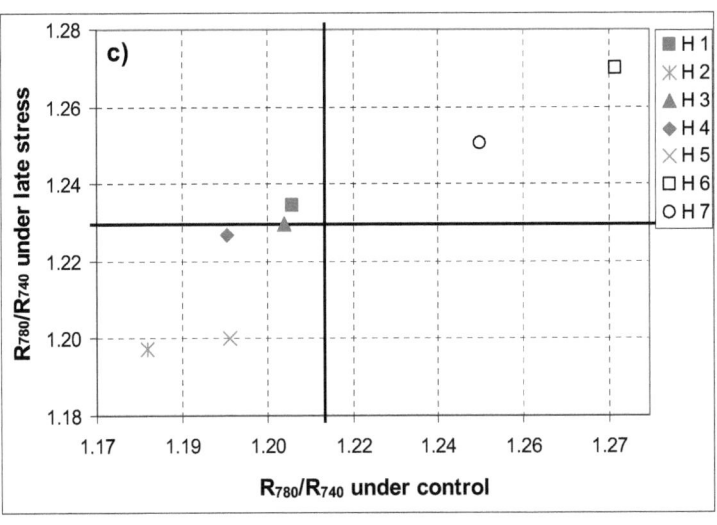

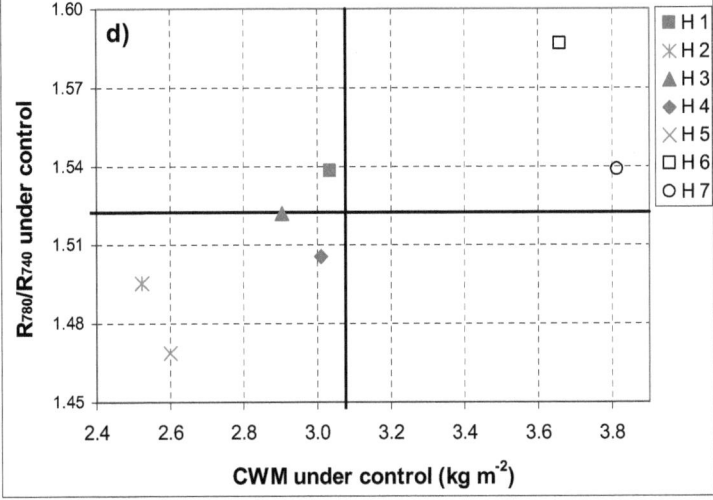

Fig. 6. Relationships between canopy water mass and spectral index R_{780}/R_{740} of hybrids for different irrigation treatments in the experimental year 2007/08. Continuous lines represent mean canopy water mass (CWM) and index value.

a) CWM of control and early vegetative stress treatments at 56 days after sowing

b) CWM of control and late vegetative stress treatments at 63 days after sowing

c) Spectral index R_{780}/R_{740} of control and late vegetative stress treatments at 42 days after sowing

d) CWM and spectral index R_{780}/R_{740} of control treatments at 63 days after sowing

4. Discussion

4.1 Soil water status

The drying out of the soil with increasing drought stress was shown through the tensiometer data, which also indicated the major rooting depth and water uptake of the maize plants. Soler et al. (2007) indicated through tensiometric measurements that drought stressed maize plants show the earliest decrease in the soil matric potential at a soil depth of 20 cm, followed by a decrease at a soil depth of 40 cm approximately 30 days after sowing, demonstrating the fast utilization of water in the upper soil layers. These results are in broad agreement with those in this study. Here, a moderate decrease in soil matric potential led to severe drought stress for the plants because of the low available soil water capacity, something that is typical for this experimental site (Camp, 1996; Neidhart, 1994). Thus, sensor measurements of the drought stressed hybrids could be assisted through measurements of the soil water status.

4.2 Detection of well-watered and drought stressed irrigation treatments

It would be very useful to be able to estimate the water content of whole canopies (%CWC) in the field using spectral reflectance data (Rollin and Milton, 1998). Given that the reflectance of maize plants is influenced by drought stress (Mistele und Schmidhalter, 2008; Schlemmer et al., 2005) and differentially so at different growth stages (Clay et al., 2006), it should be feasible to detect variation in drought stress by measuring changes in reflectance. In particular, specific wavelength bands relating to reflectance in the visible wavelength range (Hatfield et al., 2008), the red edge and the NIR band show good potential for estimating plant water status effectively (Liu et al., 2003; Yu et al., 2000), and also, for field crops, non-destructively and in real-time (Inoue et al., 1993; Seelig et al., 2009).

Our examination of selected spectral indices from the literature and newly developed for this study demonstrated that they could indeed reflect the significant influence of different irrigation treatments on the CWM of maize hybrids. Several indices in particular (R_{850}/R_{725}; R_{780}/R_{740}; R_{760}/R_{730}; R_{890}/R_{715}; $(R_{790}-R_{720})/(R_{790}+R_{720})$; R_{980}/R_{715}) showed strong relationships with CWM during the growing period, although those indices obtained from the literature had primarily been used to detect plant dry matter, nitrogen and chlorophyll. That being said, vegetation indices utilizing red and NIR wavelengths have successfully been used to detect plant water stress in the past, being highly correlated with total leaf water mass per unit ground area (Hunt and Rock, 1989). In addition, indices using the water absorption peak near 970 nm, like the WBI, could be used to detect plant water stress (Penuelas et al., 1994). However, the potential of these and similar indices extends further. For instance, our results show that the indices WI (WBI) and WI/NDVI, which have been used to estimate PWC (Penuelas et al., 1997), as well as WBI and NDVI, which have been used to estimate vegetation water status (Claudio et al., 2006) also show average relationships with CWM, with R^2 values of about 0.60. Remotely sensed leaf water indices are related to the absolute amount of water within leaves and not RWC (Seelig et al., 2008b), making them less suited to detecting CWM. Indeed literature indices measuring RWC generally showed poorly relationships with CWM. Finally, although the index R_{1300}/R_{1450} can measure leaf water content and detect the onset of leaf dehydration non-destructively and in real-time (Seelig et al., 2009), our results indicate this index to only have an average relationships with CWM (R^2 = 0.56). The cooler weather in the second experimental year led to smaller maize plants, since tropical maize reacts sensitively to chill in the vegetative stage. A lower temperature not only leads to a slower growth rate but also reduces plant size and delays flowering of tropical maize. Maize plants that experience a long period of low temperature will shorten the internodes length resulting in shorter plants and ear heights. Even though relationships in the first year were stronger and significant differences became more apparent, also in the second year close relationships were observed. Because of the smaller plants in the second year the differences in between them were small. Seen that the destructive measurements indicated only small differences and almost the same results were obtained by the non-destructive sensor-based measurements, this supports the quality of the measurements. Therefore the assumption seems to be reasonable that such a method can be extrapolated to other years, sites and environments, since the sensitivity of the sensor measurements was demonstrated to be high.

The good relationship of CWM with the newly developed spectral index R_{850}/R_{725} (R^2 = 0.83) demonstrates the excellent ability of the index to assess the water status of maize canopies. Values of the index decreased with increasing drought stress, in agreement with the results of Yu et al. (2000), who reported that drought stress impacts the physiological properties of maize significantly and leads to a decrease in the values of physiological indices with an increase of soil water stress.

Alternatively, it is possible to estimate the drought stress of maize plants through infrared thermometry (Tilling et al., 2007). Here, IR-temperature showed a good relationship with CWM (R^2 = 0.65) and was able to distinguish between the different stress treatments of the canopy, given that the more water that was available for the maize, the lower was the canopy temperature. This latter observation agrees with the results of Hunt and Rock (1989) and Peters and Evett (2007), who reported comparatively higher canopy temperatures for drought stressed plants because of the decreased stomatal conductance leading to less latent heat loss from transpiration and an increase in leaf temperature. Based on this temperature response, an array of canopy temperature sensors mounted on irrigation platforms could be used to assess the degree of water stress over large areas regularly, creating canopy temperature maps to provide a real-time snapshot of the crop status for an entire field (Hunt and Rock, 1989; Peters and Evett, 2007).

To detect the water status of maize plants through non-destructive, high-throughput sensing measurements, IR-temperature was related to the spectral index R_{850}/R_{725}, showing both a strong relationship (R^2 values between 0.65 and 0.78) and a clear differentiation of the irrigation treatments. Generally, the longer the plants experienced drought stress and the less water that was available, the lower was the index value and the higher was the temperature.

As indicated in literature, precision agriculture promises to be the next major improvement with regards to the use-efficiency of crop-production inputs, but a lack of both real-time feedback about plant status and decision support systems has hindered the practical use of site-specific irrigation (Peters and Evett, 2007). The need for such systems, however, is very real. For instance, for maize grown off-season, it is important to improve and stabilize production using supplemental irrigation (Soler et al., 2007). In such cases, measurements of canopy temperature with IR thermometry could detect spatial variations in crop water status and consequently improve management decisions by adjusting the inputs to the heterogeneous soil and growing conditions within a field. Together, this would lead to a decrease in input costs while reducing the pollution of the environment. This approach could be used in irrigated cropping systems for irrigation scheduling as well as in non-irrigated regions to reduce expensive N fertilizations on drought stressed crops (Osborne et al., 2002; Peters and Evett, 2007; Tilling et al., 2007).

In the application of sensing methods, a major improvement represented by our technique is that the sensor was mounted on a carrier vehicle and not handheld, enabling high-throughput measurements of plant parameters, which allows for a faster estimation of data on a large area. Furthermore, the exact positioning of the sensors and IR-thermometer over the maize stands with a constant angle of view and taking into account the time and duration of the measurements is considered to be advantageous. Drought stress is best measured at a rather narrow time window at or soon after high solar noon requiring a high speed and short duration of the measurements, therefore a new approach in plant breeding is vital to estimate phenotypic and physiotypic traits of drought-stressed maize hybrids. Researchers in the domain of precision farming using tractor based sensors most commonly employ a nadir view geometry, leading to sensor heights of about 9 m for maize, due to the need of at least 2.5 rows in the field of view (Major et al., 2003), or require multiple near-distance sensors that may be confronted with difficulties in matching the exact positioning of the maize rows due to the reduced foot print size. The advantage of our sensor viewing geometry is seen in the oblique and oligo view optic, reducing significantly the height of the sensor above the maize canopy and detecting simultaneously five rows. This viewing geometry allows us to obtain a large footprint of the canopies being representative of the plots and minimizes the soil influence in the field of view. Plant breeders could benefit from exactly such a highly mechanized and automated sensing precision phenotyping technique. Such methods are already used in part in precision farming, where reflectance information is obtained from tractor-mounted crop scanning devices that allow the detection of the biomass and nitrogen status of crops in real-time (Schmidhalter et al., 2001). Whereas high clearance tractors are already used in precision farming activities, the potential of high-throughput, non-destructive phenotyping platforms for analyzing and differentiating maize hybrids in breeding plots through the combination of high throughput precision phenotyping with sensors mounted on carriers and the mapping with GPS data represents an innovative and new approach demonstrated in this work.

4.3 Differentiation of hybrids

All hybrids are commercial hybrids planted in this region of Thailand. The hybrids were selected from major public and private seed agencies and have passed a wide area testing for several years before they were released to commercial uses that is why the differences among hybrids were quite small. All hybrids are high and stable in yield under normal conditions and were not selected based on a range of drought tolerance for testing in this research. The aim was to detect differences among hybrids in terms of drought tolerance in order to help breeders in future selection. The key point was that even if we detect small differences among the genotypes tested and the differences are consistent, a classification of the hybrids would be possible and this method could help in the breeding process.

The lack of correlation between CWM and grain yield can in part be ascribed to the high coefficient of variation of 18 % observed for grain yield on this location, overlapping the small differences between the hybrids. The long recovery period between the last drought stressed biomass sampling and the final harvest most likely allowed for a significant recovery as also indicated by the final yield penalty of 8 to 21 % in the drought stressed treatments, however being statistically not significant. Maize has high recovery ability from drought stress imposed in the vegetative stage being in line with previous observations at the same site (Camp, 1996; Neidhart, 1994) and supported by other studies (Pandey et al., 2000). Still the observations from this study are highly interesting because they allow to phenotype differences in the hybrids' behaviour that may potentially be useful in the further screening process. The results e.g. indicate that lower CWM of drought stressed and well-watered maize plants at the vegetative stage is not necessarily related to final yield, which could be seen as an advantage or disadvantage for the hybrids, depending on the severity of the drought stress during this period. Plants with higher CWM could have a deeper rooting system and therefore might be able to survive a longer drought period, but alternatively smaller plants may have depleted to a lesser degree the available water reserves being of relevance with continued stress. Whether the individual hybrid's recovery ability was higher or other traits such as improved harvest index or others may have contributed to this observation must be investigated in follow-up studies at later growth stages detailing further vegetative and reproductive growth parameters. The potential to differentiate vegetative growth parameters at early growth stages seems to be attractive enough to be tested in later growth stages, and potentially more close links may exist between spectrally ascertained biomass parameters in vegetative growth stages and the biomass yield of silage and energy maize. If the already demonstrated potential to discriminate hybrids can be extended to later growth stages as well, then a deeper understanding of traits contributing to the overall performance and to the final yield of drought stressed maize may become available. Further a better link of phenotypic and genotypic information may be obtained.

The tropical maize hybrids examined here influenced CWM significantly, with it being possible to classify them into three consistent groups (above, below or average performance) under control and stress environments for the investigated vegetative growing period. Similar results were reported from Edmeades et al. (1999) for biomass. Drought stressed and irrigated tropical maize show significant differences in biomass production only for cultivars with the highest and lowest total aboveground biomass. Hybrids react in different ways to limited water supplies, showing that genotypic differences exist in the vegetative responses to drought stress and management strategies, like using cultivars that are best adapted to the local environmental conditions as well as optimal irrigation strategies should be considered (Soler et al., 2007). In so doing, it is important to note that the ability to discriminate between cultivars will play a key role in screening tests of drought tolerance of maize for plant breeders (Inoue et al., 1993). Results from this work complement results from our previous study (Winterhalter et al., 2011), where spectral indices allowed to successfully estimate the aerial biomass and N uptake of seven maize hybrids subjected to different drought stress levels, with the hybrids being mostly similarly classified, as in this study investigating CWM, but reporting the amount of aerial biomass, N uptake and index values under both, control and stress environments. Although the relationships between the spectral indices and CWM were closer and higher R^2 values were achieved in this study compared to aerial biomass and N uptake reported in the previous study, a strong relationship between CWM and biomass exists demonstrating the similarity and connectivity of these important agronomical traits, which could be determined non-destructively. The assessment of phenotypic and physiotypic traits of maize hybrids with carrier based, non-destructive, high throughput reflectance and thermal measurements with accompanying GPS data opens the avenue for a new promising approach in plant breeding.

5. Conclusions

Both selected spectral indices as well as IR-temperature displayed a high correlation with CWM and were also able to discriminate between plants subjected to different drought stress levels. In addition, it was possible to classify consistently the hybrid maize plants examined in three groups (above, below or average performance) under control and stress environments. Overall, these results show that it is possible to detect the CWM of different irrigation treatments and to discriminate between groups of hybrids using high throughput phenotyping. This study, therefore, represents further evidence supporting the large potential of high throughput sensing measurements to assess the water status of crops and so assist in management decisions and especially breeding purposes (Schmidhalter, 2005). Although no consistent relationship with grain yield could be established, this method has potential to be applied in the selection process of silage or energy maize cultivars. The accurate positioning of the sensor and IR-thermometer with a constant angle of view and an oblique and oligo viewing geometry allowed obtaining a large footprint of the maize plots while minimizing the soil influence in the field of view. Reducing the time required for the measurements, being particularly vital for water status detection, further adds to a new approach in plant breeding estimating phenotypic and physiotypic traits of maize hybrids with carrier based, non-destructive, high throughput precision reflectance and thermal measurements with combined GPS data. In addition, the combination of high throughput precision phenotyping enabled through these techniques with high throughput genotyping may have the potential to provide accurate information of those plant genotypes associated with a particular phenotype and thus direct selection for the latter. Such a combined approach could lead to a breakthrough in our attempts to explain the fundamental genetic basis of drought tolerance (Campos et al., 2004; Richards et al., 2010). Further research will be required to demonstrate the potential of non-destructive high throughput assessments identifying similar or/and new traits at the vegetative and reproductive stages of maize. By that as a first step a better understanding of the effects of drought on the overall performance of maize plants may be developed, and in a second step high throughput phenotyping may also allow to identify traits being closely linked to the final biomass or grain yield.

Acknowledgement

The authors acknowledge the support of the German Federal Agency for Agriculture and Food (Project Nr. 2815303407).

References

Aparicio, N., Villegas, D., Al Araus, J. Casadesus, J., Royo, C., 2002. Relationship between growth traits and spectral vegetation indices in durum wheat. Crop Sci. 42:1547–1555.

Blum, A., 2009. Effective use of water (EUW) and not water-use efficiency (WUE) is the target of crop yield improvement under drought stress. Field Crop Res. 112:119–123.

Camp, KH., 1996. Transpiration efficiency of tropical maize (*Zea Mays* L.). Ph.D. diss. Swiss Federal Institute of Technology Zürich, Switzerland.

Campos, H., Cooper, M., Habben, J.E., Edmeades, G.O.., Schussler, J.R., 2004. Improving drought tolerance in maize: A view from industry. Field Crop Res. 90:19–34.

Chaves, M., Davies, B., 2010. Drought effects and water use efficiency: Improving crop production in dry environments. Funct. Plant Biol. 37:iii–vi.

Claudio, H.C., Cheng, Y., Fuentes, D.A., Gamon, J.A., Luo, H., Oechel, W., Qiu, H.L., Rahman, A.F., Sims, D.A., 2006. Monitoring drought effects on vegetation water content and fluxes in chaparral with the 970 nm water band index. Remote Sens. Environ. 103:304–311.

Clay, D., Kim, K., Chang, J., Clay, S., Dalsted K., 2006. Characterizing water and nitrogen stress in corn using remote sensing. Agron. J. 98:579–587.

Colombo, R., Meroni, M., Marchesi, A., Busetto, L., Rossini, M., Giardino, C., Panigada, C., 2008. Estimation of leaf and canopy water content in poplar plantations by means of hyperspectral indices and inverse modeling. Remote Sens. Environ. 112:1820–1834.

Dallon, D., 2003. Measurement of water stress: Comparison of reflectance at 970 and 1450 nm. Available at http://www.usu.edu/cpl/PDF/Water%20Stress_Dallon.pdf. Utah State University Crop Phys Lab. (verified 17 Feb. 2010).

Diker, K., Bausch, W.C., 2003. Potential use of nitrogen reflectance index to estimate plant parameters and yield of maize. Biosystems Eng. 85:437–447.

Edmeades, G.O., Bolanos, J., Chapman, S.C., Lafitte, H.R., Bänziger, M., 1999. Selection improves drought tolerance in tropical maize populations: I. Gains in Biomass, Grain Yield, and Harvest Index. Crop Sci. 39:1306–1315.

Edmeades, G.O., McMaster, G.S., White, J.W., Campos, H. 2004. Genomics and the physiologist: bridging the gap between genes and crop response. Field Crops Res. 90, 5-18.

Graeff, S., Claupein, W., 2007. Identification and discrimination of water stress in wheat leaves (*Triticum aestivum* L.) by means of reflectance measurements. Irrigation Sci. 26:61–70.

Grant, L., 1987. Diffuse and specula characteristics of leaf reflectance. Remote Sens. Environ. 22:309-322.

Hatfield, J.L., Gitelson, A.A., Schepers, J.S., Walthall, C.L., 2008. Application of spectral remote sensing for agronomic decisions. Agron. J. 100:117–131.

Hunt, E.R., Rock, B.N., 1989. Detection of changes in leaf water content using near- and middle-infrared reflectances. Remote Sens. Environ. 30:43–54.

Hu, Y., Schmidhalter, U., 2005. Drought and salinity: A comparison of their effects on mineral nutrition of plants. J. Plant Nutr. Soil Sci. 168:541-549.

Inoue, Y., Morinaga, S., Shibayama, M., 1993. Non-destructive estimation of water status of intact crop leaves based on spectral reflectance measurements. Jpn. J. Crop Sci. 62:462–469.

Liu, L., Zhao, C., Huang, W., Wang, J., 2003. Estimating winter wheat plant water content using red edge width. Int. J. Remote Sens. 25:3331–3342.

Major, D.J., Baumeister, R., Toure, A., Zhao, S., 2003. Methods of measuring and characterizing the effects of stresses on leaf and canopy signatures. ASA Spec. Publ. 66:165–175.

Mistele, B., Schmidhalter, U., 2008. Spectral measurements of the total aerial N and biomass dry weight in maize using a quadrilateral-view optic. Field Crops Res. 106: 94–103.

Mistele, B., Schmidhalter, U., 2010. A comparison of spectral reflectance and laser-induced chlorophyll fluorescence measurements to detect differences in aerial dry weight and nitrogen uptake of wheat *In* R. Khosla (ed.) 10th International Conference in Precision Agriculture. Denver, Colorado July 18-21. 2010. CD-Rom. 14p.

Neidhart, B., 1994. Morphological and physiological responses of tropical maize (*Zea mays* L.) to pre-anthesis drought. Ph.D. diss. Swiss Federal Institute of Technology Zürich, Switzerland.

Osborne, S.L., Schepers, J.S., Francis, D.D., Schlemmer, M.R., 2002. Use of spectral radiance to estimate in-season biomass and grain yield in nitrogen- and water-stressed corn. Crop Sci. 42:165–171.

Pandey, R.K., Maranville, J.W., Admou, A., 2000. Deficit irrigation and nitrogen effects on maize in a Sahelian environment. I. Grain yield and yield components. Agric. Water Manage. 46, 1–13.

Passioura, J., 2007. The drought environment: Physical, biological and agricultural perspectives. J. Expt. Bot. 113-117.

Penuelas, J., Gamon, J.A., Fredeen, A.L., Merino, J., Field, C.B., 1994. Reflectance indices associated with physiological changes in nitrogen- and water-limited sunflower leaves. Remote Sens. Environ. 48:135–146.

Penuelas, J., Pinol, J., Ogaya, R., Filella, I., 1997. Estimation of plant water concentration by the reflectance Water Index WI (R900/R970). Int. J. Remote Sens. 18:2869– 875.

Peters, R.T., Evett, S.R., 2007. Spatial and temporal analysis of crop conditions using multiple canopy temperature maps created with center-pivot-mounted infrared thermometers. ASABE 50:919-927.

Poss, J.A., Russell, W.B., Grieve, C.M., 2006. Estimating yields of salt- and water-stressed forages with remote sensing in the visible and near infrared. J. Environ. Qual. 35:1060–1071.

Richards, R.A., Rebetzke, G.J., Watt, M., Condon, A.G., Spielmeyer, W., Dolferus, R., 2010. Breeding for improved water productivity in temperate cereals: Phenotyping, quantitative trait loci, markers and the selection environment. Funct. Plant Biol. 37:85–97.

Rodriguez, D., Fitzgerald, G.J., Belford, R., Christensen, L., 2006. Detection of nitrogen deficiency in wheat from spectral reflectance indices and basic crop eco-biophysiological concepts. Aust. J. Agric. Res. 57, 781–789.

Rollin, E.M., Milton, E.J., 1998. Processing of high spectral resolution reflectance data for the retrieval of canopy water content information. Remote Sens. Environ. 65:86–92.

Schlemmer, M.R., Francis, D.D., Shanahan, J.F., Schepers, J.S., 2005. Remotely measuring chlorophyll content in corn leaves with differing nitrogen levels and relative water content. Agron. J. 97:106–112.

Schmidhalter, U., Glas, J., Heigl, R., Manhart, R., Wiesent, S., Gutser, R., Neudecker. E., 2001. Application and testing of a crop scanning instrument – field experiments with reduced crop width, tall maize plants and monitoring of cereal yield. In G. Grenier et al. (ed.) Proceedings of the 3rd European Conference on Precision Agriculture. Montpellier, France. pp. 953-958.

Schmidhalter, U., 2005. Sensing soil and plant properties by non-destructive measurements. Proceedings of the International Conference on Maize Adaption to Marginal Environments. 25th Anniversary of the Cooperation between Kasetsart University and Swiss Federal Institute of Technology. March 6-9. Nakhon Ratchasima, Thailand.

Seelig, H.D., Adams III, W.W., Hoehn, A., Stodieck, L.S., Klaus, D.M., Emery, W.J., 2008a. Extraneous variables and their influence on reflectance-based measurements of leaf water content. Irrigation Sci. 26:407–414.

Seelig, H.D., Hoehn, A., Stodieck, L.S., Klaus, D.M., Adams III, W.W., Emery, W.J., 2008b. Relations of remote sensing leaf water indices to leaf water thickness in cowpea, bean, and sugarbeet plants. Remote Sens. Environ. 112:445–455.

Seelig, H.D., Hoehn, A., Stodieck, L.S., Klaus, D.M., Adams III, W.W., Emery, W.J., 2009. Plant water parameters and the remote sensing R1300/R1450 leaf water index: controlled condition dynamics during the development of water deficit stress. Irrigation Sci. 27:357–365.

Soler, C.M.T., Hoogenboom, G., Sentelhas, P.C., Duarte, A.P., 2007. Impact of water stress on maize grown off-season in a subtropical environment. J. Agron. Crop Sci. 193:247—261.

Thoren, D., Schmidhalter, U., 2009. Nitrogen status and biomass determination of oilseed rape by laser-induced chlorophyll fluorescence. Eur. J. Agron. 30:238-242.

Tilling, A.K., O'Leary, G.J., Ferwerda, J.G., Jones, S.D., Fitzgerald, G.J., Rodriguez, D., Belford, R., 2007. Remote sensing of nitrogen and water stress in wheat. Field Crop Res. 104:77–85.

Yu, G., Miwa, T., Nakayama, K., Matsuoka, N., Kon, H., 2000. A proposal for universal formulas for estimating leaf water status of herbaceous and woody plants based on spectral reflectance properties. Plant Soil 227:47–58.

Winterhalter, L., Mistele, B., Jampatong, S., Schmidhalter, U., 2011. High throughput sensing of aerial biomass and above ground nitrogen uptake in the vegetative stage of well-watered and drought stressed tropical maize hybrids. Crop Sci. 51:1–11.

Worku, M., Bänziger, M., Schulte auf 'm Erley, G., Friesen, D., Diallo, A.O., Horst, W.J., 2007. Nitrogen uptake and utilization in contrasting nitrogen efficient tropical maize hybrids. Crop Sci. 47:519–528.

Publication III

Assessing the vertical footprint of reflectance measurements to characterize nitrogen uptake and biomass distribution in maize canopies

Highlights

- First report on influence of maize foliage vs. stem on reflectance of maize canopies

- Leaf N uptake and aerial biomass profile showed a vertical bell shape distribution

- Sensor detects N uptake of all leaf levels and separates fertilization treatments

- Vertical N uptake profile could improve nitrogen fertilization recommendations

- Useful for precision phenotyping and improving crop growth simulation models

ABSTRACT

Evaluation of phenotypic traits of crop plants on a large scale could provide important information to understand their responses to the environment. In this regard, remote sensing methods have shown much promise. However, the effect of the different contributions to spectral reflectance of the different leaf levels of plant canopies remains poorly investigated, despite their potential to improve the precision of canopy information estimation. In this study, we investigated the efficacy of sensor measurements in determining the vertical leaf nitrogen uptake of maize (*Zea mays* L.). We examined how nitrogen is distributed in the plant canopy, whether or not differences exist among fertilizer application rates, and how deep does the passive reflectance sensor meaningfully provide insight into the plant canopy. Our results, derived from either a sensor system with an oblique and multi view optic as well as SPAD measurements, indicated a convex (bulging outward) distribution of the relative chlorophyll content in maize plants. Similarly, both leaf nitrogen uptake and leaf biomass presented vertical bell shape distribution, although only the former showed qualitative differences among the fertilization treatments in the intermediate canopy leaf levels. By contrast, vertical nitrogen content presented a vertically decreasing gradient from top to bottom and one that was steeper at reduced nitrogen application. The spectral index R_{780}/R_{740} was positively and curvilinearly related ($R^2=1.00$) to the nitrogen uptake profile of the maize foliage and was able to detect the nitrogen uptake of each leaf level, even at the lowest levels. Yet, despite more than half of the total nitrogen being stored in the stem, the index values were influenced mainly by the foliage. Altogether, our results should help improve nitrogen fertilization recommendations in crop management as well as being useful in precision phenotyping and in improving in crop growth simulation models for architectural modeling.

Key words: Canopy; Corn; Crop growth simulation models; Nitrogen uptake; Phenotyping; Phenomics; Plant architecture

1. Introduction

The measurement of agronomical parameters of crop plants as a method to indicate their nutritional status as well as for the fast evaluation of phenotypic traits on a large scale provides important information to understand the response of these plants to their environment. Indeed, the accurate assessment of the chlorophyll status of plants is vital to provide nitrogen fertilization recommendations because the leaf chlorophyll concentration is determined mainly by the availability of nitrogen (Filella et al., 1995). Approximately 75% of the leaf nitrogen is integrated in the photosynthetic process and strong correlations have been identified between photosynthetic activity and leaf nitrogen content (Drouet and Bonhomme, 1999). Thus, understanding nitrogen uptake and assimilation is essential when trying to improve the nitrogen use efficiency of crops through the adjustment of nitrogen fertilizer applications (Gastal and Lemaire, 2002) and therefore for improving nitrogen management in general (Huang et al., 2011). In this regard, quantifying the chlorophyll content of plant canopies should usefully complement information on the leaf area index to help improve our understanding of crop ecophysiology, interplant competition and radiation use efficiency as well as its productivity (Ciganda et al., 2008).

Yet whereas the relationship between the chlorophyll content of leaves and the actual photosynthetic canopy area is well documented, comparatively little information is available about the vertical distribution of important plant parameters including chlorophyll, a key crop biophysical characteristic (Ciganda et al., 2008), and light and nitrogen, two crucial resources for plant development (Drouet and Bonhomme, 1999; Wang et al., 2005).

Indeed, despite the relatively constant appearance of leaves in a canopy, the canopy itself is characterized by non-uniform vertical nitrogen distribution because the leaves are exposed to different light environments (e.g. through shading), differ in age and may develop under different nitrogen supply conditions during growth (Gastal and Lemaire, 2002). A vertical gradient in the leaf nitrogen concentration is common in plant canopies, in part because nitrogen represents one of the most mobile nutrients (Wang et al., 2005). This gradient, as well as leaf nitrogen remobilization, is determined to a large extent by the local light climate during plant development, with the vertical gradient in leaf irradiance leading to a vertical gradient in leaf nitrogen content per unit area during the vegetative phase of maize (Drouet and Bonhomme, 1999). Other vertical differences in plant canopies include the bell-shaped distribution of chlorophyll content at every growth stage of maize (Ciganda et al., 2008) and a change in the total chlorophyll content in maize canopies during the growing season (Ciganda et al., 2009), as well as the slightly skewed bell-shaped function of the area-per-leaf profile in this same crop.

However, our general lack of knowledge regarding vertical distributions of plant parameters impinges on the accuracy of the models we use to predict various properties of the plants. Thus, although most models simulate canopy reflectance well, their application is limited currently to the simulation of homogeneous canopies and could lead to significant errors when applied to row crops. In all cases, the accuracy of our models depends on the realism with which the plant canopies are represented and the possibility of employing a priori knowledge on canopy properties to restrain the inversion procedure.

For instance, changes in the vertical leaf area profile of maize canopies can be interpreted biologically and might prove useful in analyzing changes in the leaf area profile inherent to growing seasons and agronomic practices (Keating and Wafula, 1992; Valentinuz and Tollenaar, 2006). It has also been suggested as an important characteristic for accurately estimating the radiation interception and canopy photosynthesis in process-based crop growth simulation models that compute dry matter accumulation from temporal integration of canopy photosynthesis based on sunlit and shaded leaf areas in the crop canopy (Valentinuz and Tollenaar, 2006). Given the importance of leaf area to crop growth, accurately predicting the former will improve the performance of crop growth models used in research and management. Here, the use of the bell-shaped distribution curve could be a promising method (Keating and Wafula, 1992). Similarly, in modeling dry matter production and grain yield in crops accurately, changes in nitrogen gradients in relation to changes in the local light climate within heterogeneous row crops like maize should be incorporated (Drouet and Bonhomme, 1999).

In a related vein, the effect of vertical footprint of plant canopies on non-destructive measurements of plant parameters has scarcely been investigated. The use of sensors for assessing biomass, nitrogen and chlorophyll levels has been increasing (Mistele and Schmidhalter, 2008; Rodriguez et al., 2006; Thoren and Schmidhalter, 2009; Tilling et al., 2007), supporting on-the go fertilizer applications

(Mistele and Schmidhalter, 2010; Schmidhalter et al., 2003) as well as enhancing high-throughput precision phenotyping in breeding (Winterhalter et al., 2011a,b; Erdle et al., 2011). In applying these techniques, however, it is important to understand exactly what the sensors are seeing and where. Yet, the vertical insight or detection footprint of the sensors is largely unknown. Do reflectance sensors collect information primarily from leaves of the top canopy or can they also penetrate the canopies of row crops such as tall maize plants and, if so, to what depth?

At least for dense canopy cover, it is well known that reflectance becomes insensitive to changes in the leaf area index (saturation effect) because the lower layers of foliage are not visible to the sensor. More generally, given that the vertical leaf layers of plant canopies contribute differently to spectral reflectance, it is important to identify their differential contributions to improve the accuracy of canopy information gained through remote sensing data (Wang et al., 2005). Here, information obtained from proximal sensing will be pivotal for referencing aircraft or satellite based sensing. In general, understanding the vertical extent of sensed information will strongly affect any agricultural decisions based on it (e.g. large-scale biomass assessment, optimized management decisions, or enhancing our understanding of plant phenotypes). For instance, fertilizer decisions are frequently based on reference values obtained from point measurements of the foliage on a specific leaf and thus may not reflect the nitrogen distribution fully (including that contained in the stems). A more complete vertical picture will also further our understanding of plant architecture and improve architectural plant modeling by providing more accurate estimations of crop biophysical characteristics with the use of realistic and accurate input variables (Casa et al., 2010).

In this work, we sought to assess the ability of sensor measurements to determine the vertical leaf nitrogen uptake and biomass distribution along the whole plant canopy in maize. Specifically, we were interested in how nitrogen and biomass are distributed in the plants; whether or not differences result from different fertilizer application rates and, if so, if they can be detected; as well as how deep into the canopy a passive reflectance based sensor reaches: does it detect the nitrogen uptake of the whole canopy or only the upper leaves? The main objective was to demonstrate the capabilities and performance of the sensor to detect the contribution of different leaf layers to the overall reflectance information. Altogether, our results will lead to a better understanding of the information to be gained from spectral measurements of plant canopies, thereby being of potential use for precision farming, enhanced phenotyping and architectural modeling.

2. Material and methods

The experiments were conducted in 2006 and 2007 at the Dürnast research station (11° 70' E, 48° 40' N, altitude 450 m) loca ted near Freising, Germany in the tertiary hills of the Bavarian Alps. The research station belongs to the Chair of Plant Nutrition at the Technical University of Munich. Average annual precipitation and temperature are 800 mm and 7.5 °C, respectively, wi th an average temperature from April to September of 13.3 °C and a sunshine durati on of 1183 h during this period. Three field experiments using maize (*Zea mays* L.) were carried out (Table 1). The field used for experiment 1 was divided into high and low yielding zones. Cultivation methods followed local technical recommendations. The seeding density was 12 plants m^{-2} and calcium ammonium nitrate was applied as a fertilizer at different rates, while no further application of P and K fertilizers was required at adequate soil P and K fertility levels. As herbicide treatment 2.5 l of Artett (Bentazon 150 g/l + Terbuthylazin 150 g/l) and 1.0 l of Motivell (Nicosulfuron 40 g/l) were used. For experiments 1 and 2, plant heights were between 160 and 220 cm, while for experiment 3 the plant heights were between 250 and 280 cm at the time of sensor measurements.

Agronomical traits of the maize canopies were measured using two different methods. First, for experiment 1, the relative chlorophyll content of the leaves was measured using the portable chlorophyll meter SPAD-502 (Minolta Camera Co., Osaka, Japan). The values obtained using this device have been shown to be well correlated with the extractable chlorophyll content and thus could be applied successfully to crop nitrogen management (Huang et al., 2011). In particular, the chlorophyll content of all maize leaf levels throughout the canopy under six different nitrogen application treatments was measured on two dates, with five replications representing five experimental units. For SPAD measurements, each leaf blade of a leaf level was measured 20 cm from the leaf tip and 1 cm from the leaf edge, an area that proved to be reliable as investigated in previous studies that included samplings all along the leaf blades (data not shown). Within this zone five measurements were obtained as an average value to represent differences among maize leaves.

Second, for experiments 2 and 3, a passive reflectance sensor mounted on a high clearance tractor (BRAUD 2714) with a measuring height of 3.1 m (above ground) was used, a setup that enables measurements of fully developed maize canopies. The sensor contained two Zeiss MMS1 silicon diode array spectrometer units, each of which can analyze the reflected radiation in 256 spectral channels with a detection range from 300 to 1000 nm and a bandwidth of 3.3 nm. Whereas the first unit was linked to a diffuser to measure the sun radiation as a reference signal, the second simultaneously measured the canopy reflectance using both an oblique and multi view optic connected to a four-in-one light fiber, thereby creating an optical mixed signal of the canopy reflection of a given area from four different directions (Mistele and Schmidhalter, 2008). The sensor detected the foliage of maize within a surface of 15 m^2 with the footprint on the soil being 3 m^2 (Fig. 1). For these experiments, canopy reflectance measurements and biomass samplings were performed alternately, while successively removing hierarchical higher leaf levels from the bottom up in the 28-m^2 area, an area larger than the sensor's field of view was chosen to avoid extraneous influences. Thus, the entire maize canopy was measured at first, followed by removal of 1-2 leaf levels, with stalks being removed in the end (Table 2). Per plot a total of about 2850 leaves were analyzed, resulting in a sample number of about 260 leaves per leaf level used for destructive measurements. For destructive measurements, the plant material was chopped and a representative subsample was weighed, oven dried at 100 °C for 24 hours and then reweighed. The dried samples were ball milled (100 µm mesh size) and analyzed for total N content using an Isotope Radio Mass Spectrometer (IRMS) combined with a preparation unit (ANCA SL 20-20; Europe Scientific, Crewe, UK) (Winterhalter et al., 2011a).

All spectral measurements used the index R_{780}/R_{740} to establish the agronomical traits of the maize canopies. This index was established previously as the optimal index for this purpose (Mistele and Schmidhalter, 2008; Winterhalter et al., 2011a). Statistical analysis of the relationships on the basis of curvilinear (quadratic) models was performed in Microsoft Excel 2003 (Microsoft Inc., Seattle, WA, USA). Non-destructive sensor measurement values were correlated with averaged destructively assessed sample values of about 260 leaves per leaf level. Leaves were sequentially numbered from the first node, therefore leaf level one was at the first detectable node.

Table 1

Details regarding the maize field experiments conducted in 2006 and 2007 at the Dürnast research station.

	Experiment 1		Experiment 2	Experiment 3
	Low	High		
Soil classification	Skeletic cambisol	Cambisol (loam)	Cambisol (silty loam)	Cambisol (silty clay loam)
Sowing date	May 4, 2006		May 4, 2006	April 25, 2007
Maize cultivar	Lacta		Lacta	Agromax
Fertilization rates (kg N ha^{-1})	140 N 80 N 0 N	200 N 120 N 0 N	180 N	300 N 180 N 120 N 0 N
Type of measurement	SPAD		Sensor & destructive	Sensor & destructive
Date of measurement (growth stage)	July 21, 2006 (BBCH 55*) August 9, 2006 (BBCH 71****)		July 24, 2006 (BBCH 63**)	July 18, 2007 (BBCH 65***)

*Middle of tassel emergence: middle of tassel begins to separate
**Male: beginning of pollen shedding; Female: tips of stigmata visible
***Male: upper and lower parts of tassel in flower; Female: stigmata fully emerged
****Beginning of grain development: kernels at blister stage, about 16% dry matter

a) top view

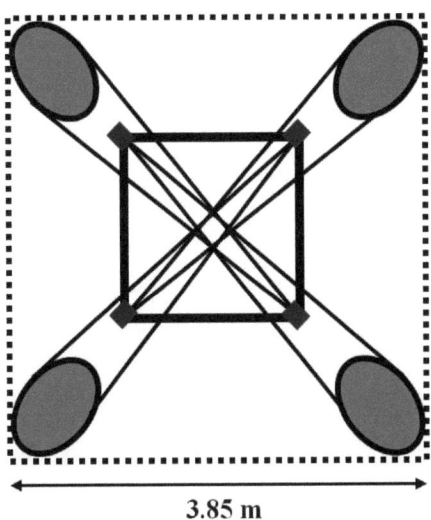

3.85 m

b) lateral view

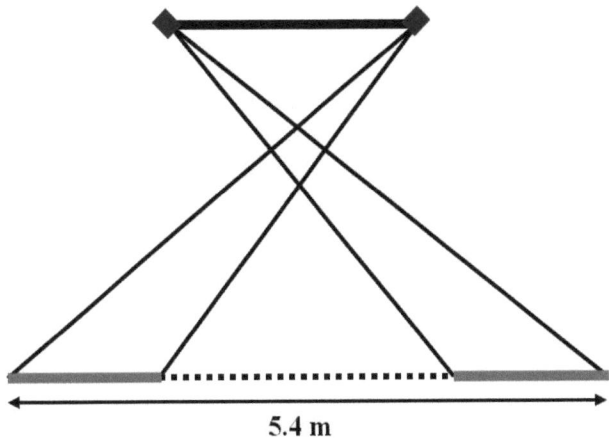

5.4 m

Fig. 1. The projected field of view with an oblique and multi viewing geometry is illustrated from the top view (a) and lateral view (b). The sensor height was 3.1 m. The area of the footprint on the soil was 3 m^2 within an area of 15 m^2, whereas totally 28 m^2 were used for accompanying destructive samplings.

Table 2

Sequence of measurements used in experiments 2 and 3, starting with reflectance measurements of the entire maize canopy, followed by the successive removal of higher leaf levels and finally of the remaining stem.

Experiment 2	Experiment 3
Original canopy	Original canopy
Leaf level 1 - 3	Leaf level 1 - 2
Leaf level 4	Leaf level 3 - 4
Leaf level 5 - 6	Leaf level 5 - 6
Leaf level 7	Leaf level 7 - 8
Leaf level 8	Leaf level 9 - 10
Leaf level 9	Flag leaf
Flag leaf	Stem
Stem	Soil
Soil	

3. Results

SPAD measurements indicated a convex (bulging outward) distribution of the relative chlorophyll content in the maize plants, with the convex shape being slightly more accentuated in the high-yielding zone (Fig. 2). The highest SPAD values appeared at or above the middle leaf levels. SPAD values generally increased with increasing nitrogen supply. This was particularly evident when compared to the 0 N ha^{-1} fertilizer (control) application; the 120 and 200 kg N ha^{-1} treatments differed only slightly at each of the two dates. Analogous SPAD values were lower at the second sampling date compared to the first.

By contrast, the nitrogen content was more evenly distributed along the different leaf levels, albeit with lower values being evident at the lower leaf levels and especially in the 0 N ha^{-1} treatment (Fig. 3). Vertical leaf biomass again followed an extreme convex distribution with the highest values at the intermediate leaf levels. Apart from the control fertilizer treatment, there was little to no differentiation in biomass among the fertilizer treatments (Fig. 4). Finally, the vertical distribution of leaf nitrogen uptake closely resembled that of leaf biomass (Fig. 5), but with a more marked difference between the fertilization treatments at the intermediate leaf levels (with increased fertilization leading to slightly higher values).

The spectral index R_{780}/R_{740} was positively and curvilinearly related to both the leaf biomass (R^2=1.00; Fig. 6) and leaf nitrogen uptake (R^2=1.00; Fig. 7) of the maize plants resulting from the successive removal of all leaf levels. With each removal event, values for each of the index R_{780}/R_{740}, leaf biomass and leaf nitrogen uptake decreased, even at the lowest leaf levels although some saturation effects are observed. In both cases, there were qualitative differences among the fertilization treatments, with increased fertilization leading to higher index values. Finally, reflectance values from the remaining stems barely differing from those of the bare soil (Table 3), with both values being noticeably lower than those obtained from the leaves.

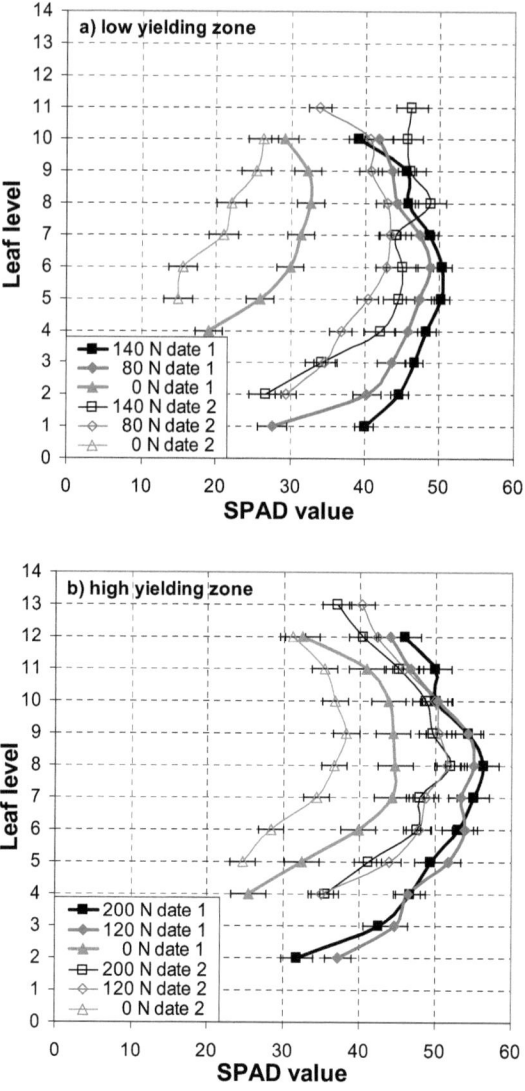

Fig. 2. Vertical distribution of the relative chlorophyll content in maize leaves as measured with SPAD on July 21, 2006, BBCH 55 (date 1, thick lines) and August 9, 2006, BBCH 71 (date 2, thin lines) for each of the low (a) and high (b) yielding zones of the experimental field under differing fertilization regimes. Missing leaf levels (e.g. levels 1 to 3) resulted from the senescence of leaves. Five replications represented five experimental units. Standard error is indicated.

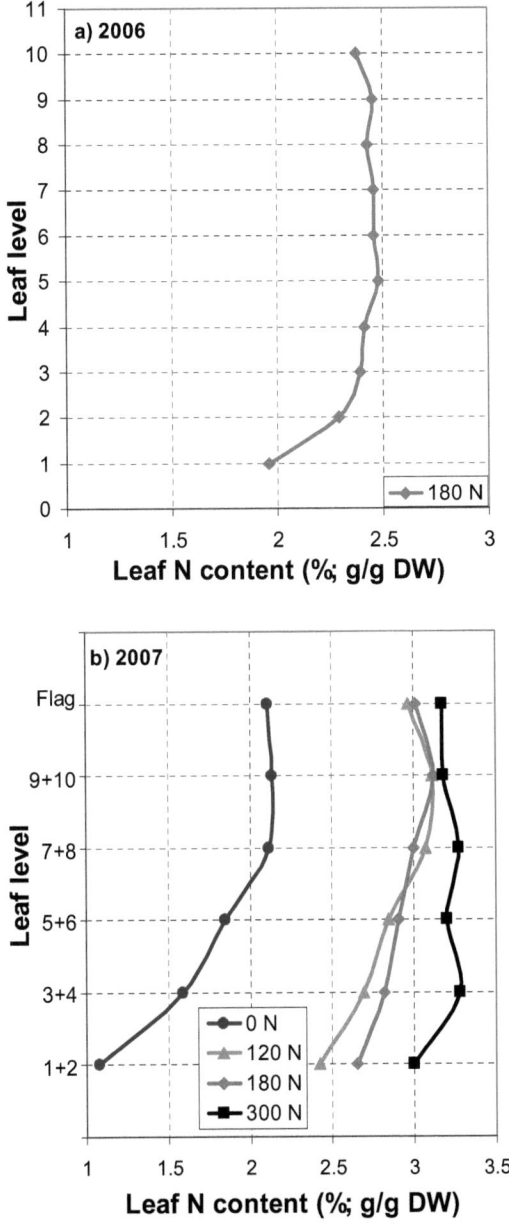

Fig. 3. Vertical distribution of the nitrogen content of maize leaves in 2006, BBCH 63 (a) and 2007, BBCH 65 (b) under different levels of nitrogen application. Each value represents an averaged sample value from about 260 leaves per leaf level from a 28-m^2 area used for the destructive measurements.

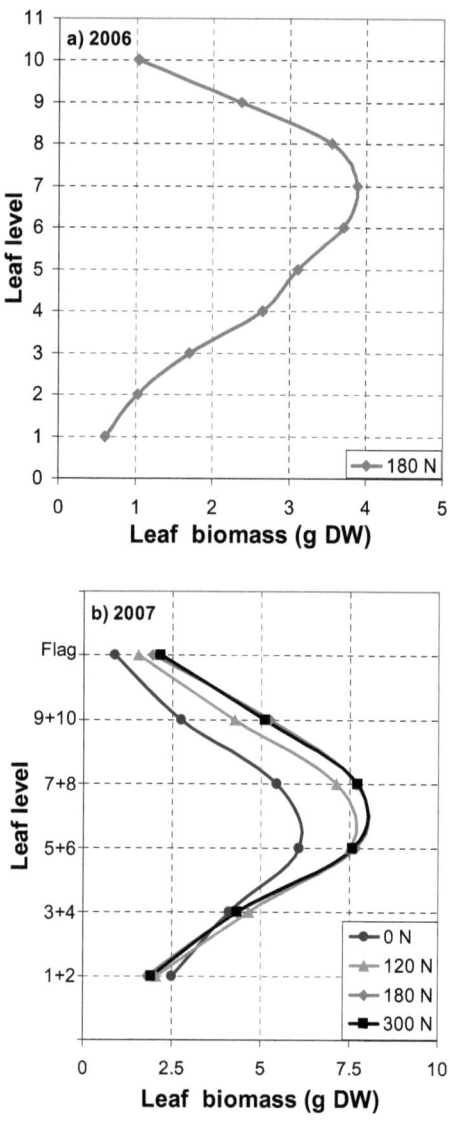

Fig. 4. Vertical leaf biomass profiles (foliage dry weight) of maize leaves in 2006, BBCH 63 (a) and 2007, BBCH 65 (b) under different levels of nitrogen application. Each value represents an averaged sample value from about 260 leaves per leaf level from a 28-m^2 area used for the destructive measurements. Foliage dry weight per square meter can be obtained by multiplying values with plants per square meter.

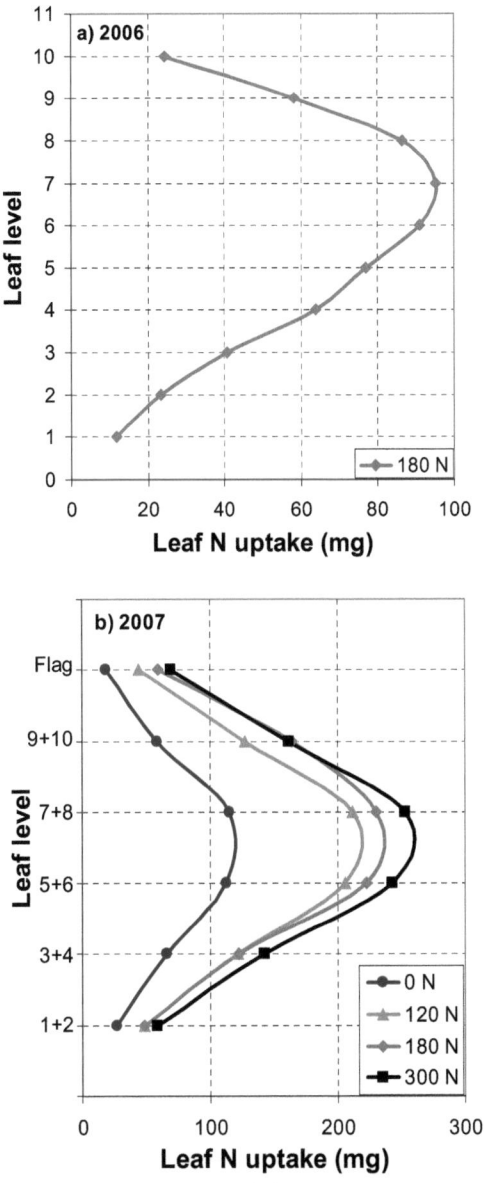

Fig. 5. Vertical leaf nitrogen uptake profiles of maize leaves in 2006, BBCH 63 (a) and 2007, BBCH 65 (b) under different levels of nitrogen application. Each value represents an averaged sample value from about 260 leaves per leaf level from a 28-m^2 area used for the destructive measurements. Nitrogen uptake per square meter can be obtained by multiplying values with plants per square meter.

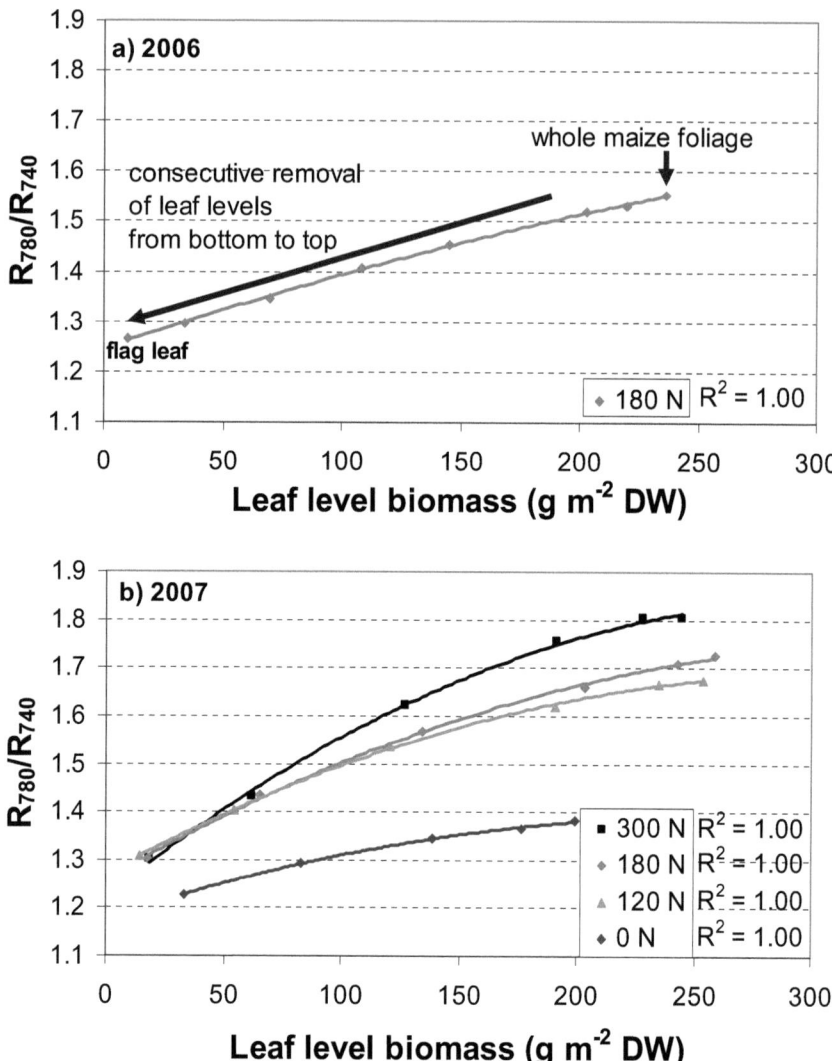

Fig. 6. Relationship of the spectral index R_{780}/R_{740} with the leaf biomass of maize canopies (foliage dry weight) resulting from the successive removal of all leaf levels in 2006, BBCH 63 (a) and 2007, BBCH 65 (b) under different levels of nitrogen application. Each value represents an averaged sample value from about 260 leaves per leaf level from a 28-m^2 area used for the destructive measurements.

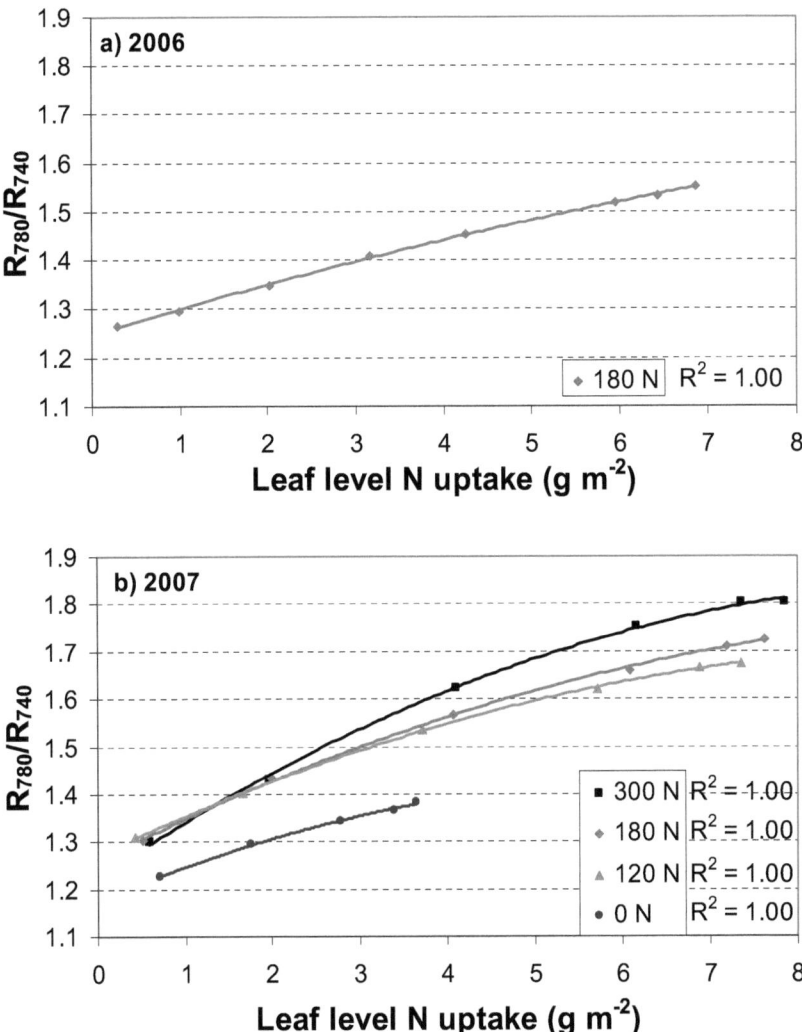

Fig. 7. Relationship of the spectral index R_{780}/R_{740} with leaf nitrogen uptake of maize canopies resulting from the successive removal of all leaf levels in 2006, BBCH 63 (a) and 2007, BBCH 65 (b) under different levels of nitrogen application. Each value represents an averaged sample from about 260 leaves per leaf level from a 28-m^2 area used for the destructive measurements.

Table 3

Spectral index R_{780}/R_{740} values of the whole maize canopy, the stem after the removal of all leaves, and the soil in 2007 (BBCH 65). Nitrogen contained in the leaves and the stem as well as the respective biomass of either is also indicated.

		R_{780}/R_{740}	N uptake (g m^{-2})	Aerial biomass (g m^{-2})
300 N	Leaves	1.80	7.86	244.51
	Stem	1.24	9.81	594.50
	Soil	1.23	0	0
180 N	Leaves	1.73	7.63	259.09
	Stem	1.26	8.84	610.39
	Soil	1.20	0	0
120 N	Leaves	1.68	7.36	253.81
	Stem	1.26	8.50	604.13
	Soil	1.20	0	0

4. Discussion

Similar to the results of Ciganda et al. (2008), our analyses showed that the vertical chlorophyll content distribution of maize follows a bell-shaped curve with the highest values at the intermediate leaf levels. This is in contrast to winter wheat, where the vertical chlorophyll distribution shows a decreasing trend from the top to the ground surface (Huang et al., 2011). The SPAD values generally increased with higher nitrogen supplies, but were similar for the 120 and 200 kg N ha^{-1} treatments in the high yielding region, possibly indicating that the maximal chlorophyll content had already been reached in both maize canopies. Similarly shaped curves were also found for leaf biomass and leaf nitrogen uptake, albeit qualitative differences among the fertilization treatments at the intermediate leaf levels were present for nitrogen uptake only. To our knowledge, there is no information about these parameters in the literature, merely that the area-per-leaf profile of maize canopies follows a slightly skewed bell-shaped function (Keating and Wafula, 1992; Valentinuz and Tollenaar, 2006). By contrast, the vertical nitrogen content of the maize plants was more evenly distributed, with steeper gradients from top to bottom at the lower fertilization rates. Similar results have been reported in literature: a vertical gradient in the leaf nitrogen concentration is widespread in crop canopies (Wang et al., 2005), with a vertical gradient in leaf nitrogen content per unit area in the vegetative phase of maize having also been shown (Drouet and Bonhomme, 1999).

To improve the nitrogen use efficiency by adjusting the amount of nitrogen fertilizer applied to crops, it is important to advance the understanding of nitrogen uptake and assimilation (Gastal and Lemaire, 2002). Since the nitrogen partitioning during maize development is rarely considered in crop models, the precise modeling of dry-matter production could be possible with the information of the nitrogen profile (Drouet and Bonhomme, 1999) and improve the performance of crop growth models used in research and management (Keating and Wafula, 1992; Valentinuz and Tollenaar, 2006).

The spectral index R_{780}/R_{740}, which has been used in previous studies to relate agronomical traits of maize canopies with spectral reflectance (Mistele and Schmidhalter, 2008; Winterhalter et al., 2011a,b), was positively and curvilinearly related to leaf biomass and leaf nitrogen uptake of maize and was also able to show differences among the fertilization treatments. Successive removal of the leaf levels yielded decreased values of the index, even at the lowest leaf levels. However, the small saturation effect observed indicates that the sensor is underestimating the leaf biomass and leaf nitrogen uptake of the lowest leaf levels. This is in stark contrast to the results of Casa et al. (2010) who reported reflectance becoming increasingly insensitive to changes in the leaf area index because the lower layers of foliage were not visible at all. The fact that only a small saturation effect was present in this study, especially with high fertilization rates during the late growing period, might prove useful in enhancing high-throughput phenotyping (Winterhalter et al., 2011a,b). It also means that the sensor system would certainly detect the nitrogen uptake in an earlier growing stage more accurately (because there is less overlaying foliage present), potentially supporting on-the-go fertilizer applications (Mistele and Schmidhalter, 2010; Schmidhalter et al., 2003).

Finally, although more than half of the total nitrogen in maize was stored in the stem, reflectance index values were influenced mainly by the foliage, with values from remaining stems and the bare soil hardly differing. Again, to the best of our knowledge, this is the first report on the impact of maize stem (versus the foliage) on the proportion of the reflectance of maize canopies.

Overall our results consistently show the same general relationship over years and at different fertilization treatments with the sensor being able to track information from the lowest leaf layers, however with slightly varying degree due to differences in the canopy structure. However, for investigating the vertical footprint size of the sensor a different approach and experimental design had to be chosen instead of using replicated point measurements of single leaves. In our study we used an average sample of hundreds of leaves for each data point in the field of view. Since this was very laborious requiring a high manpower being involved in the sampling process and keeping in mind the likely influence of the zenith angle on the sensor measurements, therefore we had to choose an alternative approach that did not allow to encompass replicated measurements. However, we argue that the large sampling number involved allows for sound conclusions that are further corroborated and supported by the results from the other measurements conducted over two different years. Clearly removal of single leaves as replicated point measurements would not have allowed arriving at the same information, since all leaves within the sensor's field of view had to be removed and replicated measurements were not possible for time reasons as mentioned above. The evaluation of a large surface with a huge number of leaves, over several years, fertilization treatments and varieties offers however an excellent insight in the vertical footprint size, seen also that the same principle was demonstrated in every experimental unit. We did not expect a similarity over seasons, since conditions vary over years, e.g. meteorological conditions, growth stages of plants, soil nitrogen supply, but it was clearly shown that the sensor detected the aerial biomass and nitrogen uptake of every leaf level. The values of the destructive as well as non-destructive measurements varied over the years, but the main findings were analogous, since the relativity of the spectral measurements was not affected by differences between seasons.

5. Conclusions

In this work, we presented the vertical distributions of chlorophyll, leaf biomass, nitrogen content and uptake in maize canopies as a prelude to determining the vertical insight or footprint of a passive reflectance sensor. Many of our results, to the best of our knowledge, represent the first data available in this context. Our results show that the sensor could indeed detect the leaf biomass and leaf nitrogen uptake of the lowest leaf levels, albeit with a slight underestimation, as well as to show differences among the fertilization treatments, supporting the possibility of incorporating such information to improve nitrogen fertilization strategies in crop management. The information of the vertical nitrogen profile could also be used in crop growth simulation models used for research purposes. A more accurate estimation of crop biophysical characteristics should be possible by improving the realism of models used for the inversion with remotely sensed data (Casa et al., 2010). Future research analyzing different maize hybrids would be a further valuable contribution for extending high-throughput precision phenotyping.

Acknowledgements

The authors acknowledge the support of the German Federal Agency for Agriculture and Food (Project Nr. 2815303407).

References

Casa, R., Baret, F., Buis, S., Lopez-Lozano, R., Pascucci, S., Palombo, A., Jones, H.G., 2010. Estimation of maize canopy properties from remote sensing by inversion of 1-D and 4-D models. Precision Agric 11, 319–334.

Ciganda, V., Gitelson, A., Schepers, J., 2008. Vertical profile and temporal variation of chlorophyll in maize canopy: Quantitative "Crop Vigor" indicator by means of reflectance-based techniques. Agron. J. 100, 1409–1417.

Ciganda, V., Gitelson, A., Schepers, J., 2009. Non-destructive determination of maize leaf and canopy chlorophyll content. J. Plant Physiol. 166, 157–167.

Erdle, K., Mistele, B., Schmidhalter, U., 2011. Comparison of active and passive spectral sensors in discriminating biomass parameters and nitrogen status in wheat cultivars. Field Crops Res. 124, 74-84.

Drouet J.-L., Bonhomme, R., 1999. Do variations in local leaf irradiance explain changes to leaf nitrogen within row maize canopies? Ann. Bot.-London 84, 61–69.

Filella, I., Serrano, L., Serra, J., Penuelas J., 1995. Evaluating wheat nitrogen status with canopy reflectance indices and discriminant analysis. Crop Sci. 35, 1400–1405.

Gastal F., Lemaire G., 2002. N uptake and distribution in crops: an agronomical and ecophysiological perspective. J. Exp. Bot. 53, 789–799.

Huang, W., Wang, Z., Huang, L., Lamb, D.W., Ma, Z., Zhang, J., Wang, J., Zhao. C., 2011. Estimation of vertical distribution of chlorophyll concentration by bi-directional canopy reflectance spectra in winter wheat. Precision Agric 12, 165–178.

Keating B.A., Wafula, B.M., 1992. Modelling the fully expanded area of maize leaves. Field Crops Res. 29, 163–176.

Mistele, B., Schmidhalter, U., 2008. Spectral measurements of the total aerial N and biomass dry weight in maize using a quadrilateral-view optic. Field Crops Res. 106, 94–103.

Mistele, B., Schmidhalter, U., 2010. Tractor-based quadrilateral spectral reflectance measurements to detect biomass and total aerial nitrogen in winter wheat. Agron. J. 102, 499–506.

Rodriguez, D., Fitzgerald, G., Belford, R., Christensen, L., 2006. Detection of nitrogen deficiency in wheat from spectral reflectance indices and basic crop eco-biophysiological concepts. Aust. J. Agric. Res. 57, 781–789.

Schmidhalter, U., Jungert, S., Bredemeier, C., Gutser, R., Manhart, R., Mistele, B., Gerl, G., 2003. Field-scale validation of a tractor based multispectral crop scanner to determine biomass and nitrogen uptake of winter wheat. In: Stafford, J., Werner, A. (Eds.), Precision Agriculture: Proceedings of the 4th European Conference on Precision Agriculture. Academic Publishers, Wageningen, pp. 615–619.

Thoren, D., Schmidhalter, U., 2009. Nitrogen status and biomass determination of oilseed rape by laser-induced chlorophyll fluorescence. Eur. J. Agron. 30, 238–242.

Tilling, A., O'Leary, G., Ferwerda, J., Jones, S., Fitzgerald, G., Rodriguez, D., Belford, R., 2007. Remote sensing of nitrogen and water stress in wheat. Field Crops Res. 104, 77–85.

Valentinuz, O.R., Tollenaar, M., 2006. Effect of genotype, nitrogen, plant density, and row spacing on the area-per-leaf profile in maize. Agron. J. 98, 94–99.

Wang, Z., Wang, J., Zhao, C., Zhao, M., Huang, W., Wang, C., 2005. Vertical distribution of nitrogen in different layers of leaf and stem and their relationship with grain quality of winter wheat. J. Plant Nutr. 28, 73–91.

Winterhalter, L., Mistele, B., Jampatong, S., Schmidhalter, U., 2011a. High throughput sensing of aerial biomass and above ground nitrogen uptake in the vegetative stage of well-watered and drought stressed tropical maize hybrids. Crop Sci. 51, 479–489.

Winterhalter, L., Mistele, B., Jampatong, S., Schmidhalter, U., 2011b. High throughput phenotyping of canopy water mass and canopy temperature in well-watered and drought stressed tropical maize hybrids in the vegetative stage. Eur. J. Agron. 35, 22-32.

i want morebooks!

Buy your books fast and straightforward online - at one of world's fastest growing online book stores! Environmentally sound due to Print-on-Demand technologies.

Buy your books online at
www.get-morebooks.com

Kaufen Sie Ihre Bücher schnell und unkompliziert online – auf einer der am schnellsten wachsenden Buchhandelsplattformen weltweit! Dank Print-On-Demand umwelt- und ressourcenschonend produziert.

Bücher schneller online kaufen
www.morebooks.de

VDM Verlagsservicegesellschaft mbH
Heinrich-Böcking-Str. 6-8 Telefon: +49 681 3720 174 info@vdm-vsg.de
D - 66121 Saarbrücken Telefax: +49 681 3720 1749 www.vdm-vsg.de

Printed by Books on Demand GmbH, Norderstedt / Germany